The Triple Challenge for Europe

The Triple Challenge for Europe

Economic Development, Climate Change, and Governance

Edited by
Jan Fagerberg, Staffan Laestadius, and Ben R. Martin

OXFORD
UNIVERSITY PRESS

OXFORD
UNIVERSITY PRESS

Great Clarendon Street, Oxford, OX2 6DP,
United Kingdom

Oxford University Press is a department of the University of Oxford.
It furthers the University's objective of excellence in research, scholarship, and education by publishing worldwide. Oxford is a registered trade mark of Oxford University Press in the UK and in certain other countries

First Edition published in 2015

Published in the United States of America by Oxford University Press
198 Madison Avenue, New York, NY 10016, United States of America

British Library Cataloguing in Publication Data
Data available

Library of Congress Control Number: 2015933913

ISBN 978–0–19–874741–3

Acknowledgements

The point of departure for this book was discussions during a workshop on 'The Challenge for Europe in a New Age' in Ålborg, Denmark, 14–15 March 2013, during which early drafts of most of the chapters in this volume were presented. The editors wish to express their gratitude to the Obel Family Foundation (www.obel.com/en), the Department for Business and Management at Aalborg University (http://www2.business.aau.dk), and the IKE Research Group (www.ike.aau.dk) for their support of the initial workshop as well as during the process that followed. Adriana-Loredana Lifa provided invaluable assistance in the process of assembling and editing the final manuscript. At Oxford University Press, David Musson and Clare Kennedy were helpful and inspiring partners.

Contents

List of Figures

List of Tables

List of Boxes

List of Abbreviations

AL	Albania
AT	Austria
BA	Bosnia and Herzegovina
BE	Belgium
BG	Bulgaria
BY	Belarus
CY	Cyprus
CZ	Czech Republic
DE	Germany
DK	Denmark
EE	Estonia
EL	Greece
ES	Spain
FI	Finland
FR	France
HR	Croatia
HU	Hungary
IE	Ireland
IT	Italy
KZ	Kazakhstan
LT	Lithuania
LU	Luxembourg
LV	Latvia
ME	Montenegro
MK	Macedonia
MT	Malta
NL	Netherlands
PL	Poland

PT	Portugal
RO	Romania
RS	Serbia
RU	Russia
SE	Sweden
SI	Slovenia
SK	Slovakia
TR	Turkey
UA	Ukraine
UK	United Kingdom
XK	Kosovo

List of Contributors

Iain Begg, Professorial Research Fellow at the European Institute, London School of Economics and Political Science.

Jan Fagerberg, Professor at the University of Oslo and Aalborg University.

Frank W. Geels, Professor of System Innovation and Sustainability at the Sustainable Consumption Institute, University of Manchester.

Staffan Jacobsson, Professor in Environmental System Analysis, Energy and Environment, Chalmers University of Technology.

Staffan Laestadius, Professor of Industrial Dynamics, KTH Royal Institute of Technology.

Michael A. Landesmann, Scientific Director of The Vienna Institute for International Economic Studies (wiiw) and Professor of Economics, Johannes Kepler University.

Volkmar Lauber, formerly Professor of Comparative Politics at the University of Salzburg.

Rasmus Lema, Assistant Professor, Aalborg University.

Ben R. Martin, Professor of Science and Technology Policy Studies in SPRU, University of Sussex.

Mariana Mazzucato, Professor of Economics of Innovation at SPRU, University of Sussex.

Carlota Perez, Researcher, Professor of Technology and Development, London School of Economics and Political Science.

Hubert Schmitz, Professorial Fellow, Institute of Development Studies.

Bart Verspagen, Director of UNU-MERIT and Director-Dean of the Maastricht Graduate School of Governance, Maastricht University.

1

Introduction

The Triple Challenge for Europe

Jan Fagerberg, Staffan Laestadius, and Ben R. Martin

1.1. Introduction

This book argues that Europe today is confronted by fundamental changes in its external environment as well as internally, giving rise to several daunting policy challenges. First, there is *the economic challenge* manifest in slow growth or even stagnation in many developed countries. This is related to a number of factors. One is the globalization of the world economy and the rise of new economic powers in Asia and elsewhere, posing a threat to the competiveness of many national industries throughout the developed world while at the same time opening up new markets and opportunities. A second is the extensive and long-lasting impact of the 2007–8 financial crisis and the measures taken to counter its effects. In recent years, economic stagnation has been especially pronounced in Europe, where it has led among other things to increased differences between rich and poor countries and to very high levels of unemployment (especially among the young) in large parts of the continent. Third, there are underlying and persistent structural problems in certain countries that need to be urgently addressed, as we shall see in Chapters 2 and 3. Consequently, the nature and scale of the economic challenge varies considerably across countries. Austerity programmes currently under way in a number of European countries are unlikely to provide a satisfactory solution to these problems. Rather, as we shall see in what follows, what is required is a continuous upgrading of skills, technologies, and industrial structures in the countries concerned, and long-term policies—backed up with substantial investment—to make all this possible.

Second, there is *the challenge posed by the climate crisis*, the solution of which requires nothing less than a fundamental transformation from carbon-based growth to a new, sustainable economy. This may be technologically feasible but is nevertheless extraordinarily difficult to bring about as it challenges fundamental beliefs regarding the economy and society and our very notion of 'progress'. Yet without a successful transformation to a sustainable economy, future generations will be in dire straits. Like the economic challenge, the climate challenge points to the need for long-term policies supporting the development of skills, innovation, and structural change in Europe.

The third challenge concerns *the governance crisis now facing Europe* and the difficulties this poses for policy-making and implementation. These difficulties were already becoming apparent before 2008, but the financial crisis has made them much more severe. A European Union that is struggling with dwindling popular support for its institutions is finding it difficult to rally its citizens behind adequate responses to such long-term problems. That task has been made all the more difficult by the hegemonic influence of neo-liberal ideology over the last two decades, with its blind faith in the supremacy of 'market solutions' and its constant belittling of any efforts by 'the state'.

This book argues that a completely new policy stance is required, one that simultaneously addresses the challenges brought about by economic stagnation, by climate change, and by the governance crisis. Challenges, while often daunting, are also opportunities for doing things differently, and may call forth—and indeed even require—innovation and entrepreneurship at various levels of society, not only in private enterprises but also in the public sector and among policy-makers, for their successful resolution.

The purpose of this book is to contribute to an informed debate in Europe and elsewhere about these issues. It is our hope that the book will be of interest not only to specialists but also to decision-makers in both public and private sectors, to students, and to concerned citizens more generally. In addition to this introductory chapter, the book contains three sections focusing on the economic challenge, the climate challenge, and the policy and governance challenge, respectively. This chapter summarizes the main issues and arguments of the book, and sets out the conclusions with respect to how these challenges can be dealt with more effectively in order to arrive at a more economically secure, environmentally sustainable, and well governed Europe.

1.2. The Economic Challenge for Europe

It is important to remember that, over the longer term, European economic integration has delivered very substantial benefits to Europe's citizens. For example, during the first two or three decades of integration efforts in

(Western) Europe the economy grew very fast, and the gap in productivity and income vis-à-vis the world's technologically and economically leading country, the United States, was considerably reduced (Abramovitz, 1994a, b). The diffusion of technologies and innovations, including ways to organize business and consumption patterns, that were pioneered in the US before the war but only applied in Europe on a major scale from the 1950s onwards, played an important role in the rapid catch-up process by Western European countries during the decades that followed. The emerging economic integration of Europe, by providing a sufficiently large and growing market, was instrumental in supporting the growth of the new industries of the time.

The European Union has also been successful in supporting transitions from authoritarian regimes to democracy in large parts of Europe. This happened first from the mid-1970s onwards when the fascist dictatorships in Southern Europe were swept away following popular uprisings, and once more—and on a much larger scale—in the 1990s and 2000s following the disintegration of the former Soviet Union. The gradual integration of the countries in Eastern Europe, followed by substantial inflows of investment from the rest of Europe, led to very rapid growth in the new member countries and markedly reduced differences in productivity and income across Europe as a whole, as shown in Chapter 2 by Fagerberg and Verspagen.

Around the turn of the millennium, several initiatives were taken to sustain the positive dynamics of the previous decades in the expectation that this would lead to a further narrowing of the gap in GDP per capita between the US and Europe. For example, at the EU summits in Lisbon and Barcelona in 2000 and 2002, Member States agreed on the goal of making Europe 'the most competitive and dynamic knowledge-based economy in the world capable of sustainable economic growth with more and better jobs and greater social cohesion' by increasing R&D investments (as a share of GDP) to a level superior to that of the USA by the end of the decade.[1] Moreover, a common European currency, the euro, was introduced in 2002 as part of the strategy to further deepen European integration and spur economic growth. These initiatives appear not to have had a very great effect on Europe's growth, however, as shown in Figure 1.1, which traces the development of GDP per capita from 2003 onwards for four groups of European countries (and for comparison the United States), measured in constant, purchasing power parity (PPP) adjusted, US dollars.[2]

[1] See <www.consilium.europa.eu/en/uedocs/cms_data/docs/pressdata/en/ec/00100-r1.en0.htm> (accessed Jan. 2015). However, this was never much more than an optimistic aspiration and, with the exception of a few countries in Northern Europe, little progress has been made towards achieving this aim.

[2] These country groups, which are also used in Ch. 2 of this volume, are mostly self-explanatory, except perhaps for the residual category of 'other Europe', which consists of countries in the Middle

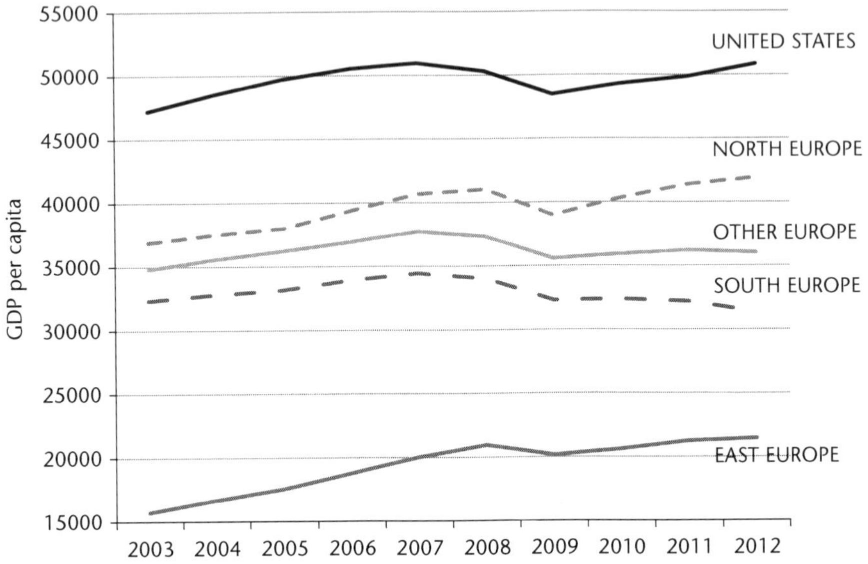

Figure 1.1. GDP per capita, constant US$, 2003–12

Note: Authors' calculations of population-weighted GDP per capita, in constant US$ (PPP) at 2013 price levels, for selected country groups, based on data from the World Bank (<http://data.worldbank.org/indicator/NY.GDP.PCAP.PP.CD> and <http://data.worldbank.org/indicator/SP.POP.TOTL>, accessed Sept. 2014)

In fact, in 2012 GDP per capita in the European Union was 65 per cent of the US level, exactly the same as ten years earlier—in other words, no catch-up vis-à-vis the US took place during this period. Among the European countries, only the new members from the East of Europe managed to substantially reduce the productivity gap vis-à-vis the US, rising from 33 per cent to 42 per cent of the US level between 2003 and 2008, after which the catch-up of Eastern Europe came to an abrupt halt.[3] The Northern European countries recorded a slight increase over the same period, from 78 per cent in 2003 to 82 per cent in 2008, with no gains thereafter. However, in Southern Europe GDP per capita dropped from 69 per cent to 62 per cent of the US level between 2003 and 2012, and the average GDP per capita level there in 2012 was

and West of Europe (Belgium, France, UK, Ireland and some smaller countries not included elsewhere). Germany is included in the 'North' together with the Nordic countries and the Netherlands. The Southern countries are Greece, Italy, Portugal, and Spain, while the Eastern group consists of the eleven member countries formerly part of, or dominated by, the Soviet Union. See Ch. 2 for further details.

[3] By 2012, only about half of the Eastern member countries had managed to recover to pre-crisis levels of GDP per capita. The somewhat better performance of Eastern Europe as a group (see Figure 1.1) owes much to Poland, the most populous country in the region, which exhibited a markedly better performance than other European economies after the financial crisis struck in 2007–8.

approximately 10 per cent below the pre-crisis level (see Figure 1.1). Thus, rather than the convergence in GDP per capita that characterized Europe during the previous decade, the years after 2007–8 witnessed a process of *divergence*, with several countries, particularly in the Southern part of the continent, falling further behind economically.

Should we be concerned about these developments? Yes—and to see why, consider Figure 1.2, which shows unemployment rates for young adults aged 20–24 in different parts of Europe since the crisis struck (see also Chapter 3 in this volume).

The slow growth and increasing divergence that characterize the European economy are clearly mirrored in Figure 1.2: apart from in Northern Europe, youth unemployment has been on the increase everywhere. In Southern Europe, the level of youth unemployment has tripled compared to the situation before the financial crisis, and currently close to half of all young adults seeking a job there cannot find one. In Eastern Europe, the level of youth unemployment has doubled over the same period, and the situation is also worrying in other countries such as France. This is, of course, just the opposite of the goal stated more than a decade ago of 'sustainable economic growth with more and better jobs and greater social cohesion'. If this situation is not reversed quickly, large numbers of young people in Europe risk being permanently marginalized—i.e. unemployed and living in relative poverty—the

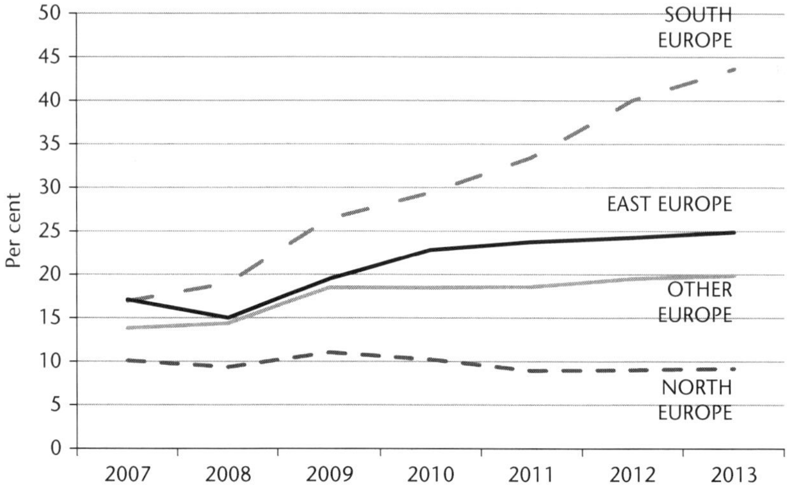

Figure 1.2. Youth unemployment (20–24 years)

Note: Authors' calculations of population-weighted unemployment rates for selected country groups based on data from Eurostat (<http://appsso.eurostat.ec.europa.eu/[yth_empl_090]>, accessed Sept. 2014) for unemployment and the World Bank (<http://data.worldbank.org/indicator/SP.POP.TOTL>, accessed Sept. 2014) for population

social, economic, and political consequences of which are likely to be highly adverse for Europe's future.

Why is Europe's growth performance so disappointing? Economic theory tells us that growth in GDP per capita is to a considerable extent brought about by innovation and diffusion of technology (Fagerberg, 1994). This process also opens up the possibility for actors who are not at the innovation frontier to tap into the growth potential of innovation by exploiting the scope for diffusion (or 'spill-overs'). According to economic historians such as Gerschenkron (1962) and Abramovitz (1986), the further behind a country lies, the larger the potential for 'catch-up' in productivity vis-à-vis the technologically leading nations. Hence, from this perspective one might have expected the countries in the East and South of Europe to have grown substantially faster than the richer countries in the Northern part of the continent, the opposite of what has been observed recently. However, the leading theoretician on catching-up processes, Moses Abramovitz, was at pains to stress the difference between a *potential* for catch-up and its *realization* (Abramovitz, 1994a, b). The latter, he argued, also depends on how favourable the external regional environment is, the nature of the technological and structural changes that occur globally, and how well placed (or adapted) a country is to profit from these changes. Hence, to understand the reasons behind the present stagnation and divergence in Europe, it is necessary to examine the factors influencing these developments in more detail.

The economic changes that have taken place in Europe during the last two or three decades have occurred within an international context characterized by globalization. The ICT revolution, combined with innovations in transport technology, provided the basis for the development of new business models based on the coordination of activities at a global scale, thereby separating the different stages of production in order to exploit the various advantages that different locations across the globe may possess. The gradual inclusion of China in the global capitalist economy, adding hundreds of millions of lower-paid manufacturing workers to the global labour pool, and effectively transforming China into 'the factory of the world', provided a substantial boost to this process. As a result, the Chinese economy has grown very fast and, as Chapter 2 reveals, similar—although somewhat less spectacular—developments have taken place in other developing nations. This process, while hugely beneficial for the hundreds of millions lifted out of poverty (and for the global economy as a whole), also poses a challenge, however, because it tends to undermine the competitive position of established industries throughout the developed world, especially in low-skill, labour-intensive manufacturing processes. Indeed, it even affects newer industries such as those in 'green' sectors, as Chapter 5 by Schmitz and Lema demonstrates. However, globalization also offers new opportunities for exports of advanced products to the rapidly growing markets in China and elsewhere, as countries such as Germany and Sweden have recognized and been keen to exploit. The evidence considered in Chapters 2 and 3 suggests that the effects of globalization on the growth performance of different parts of Europe have been very uneven. While the advanced economies in the North of Europe have to a certain extent managed to adapt to the changing competitive conditions and taken advantage of the opportunities opened up by globalization, substantially increasing their exports as a percentage of GDP, countries in the Southern part of the continent (and some in the East) have generally failed to do so. Indeed, their low-tech manufacturing sectors appear to have been particularly vulnerable to the consequences of globalization. Thus, they have become losers in the globalization process.

Europe's growth performance is also influenced by how European integration has developed, and by the strategies pursued by leading economies in the Union, especially Germany. The introduction of the euro in 2002 meant that the economies of the Eurozone became more interdependent. A natural consequence of this would have been a greater degree of coordination of economic policies among the participating countries. But the Eurozone countries continued to shape their economic policies largely based on domestic considerations, effectively disregarding the consequences for other countries and the wider Eurozone. This clearly illustrates the governance and policy challenge confronting Europe. For example, Germany, following the costly reunification with former East Germany, decided to restrain growth in wages and domestic demand in order to boost the competitiveness of its industry and to run a trade surplus vis-à-vis the rest of the world. However, this policy implied that other, less competitive members of the Eurozone, with less scope for growing through increasing exports, would have to practice austerity as well (if increasing trade deficits were to be avoided), with rising unemployment an almost inevitable consequence. Initially, several countries in the South shied away from austerity, leading (as shown in Chapters 2 and 3) to increasing deficits and rising foreign indebtedness, a situation which clearly was not sustainable. However, eventually the financial crisis brought governments in different parts of Europe together under the umbrella of austerity, leading to slow growth, rising unemployment (especially in the South), and increasing divergence in the Union as whole.

The slow growth and increasing divergence in Europe can therefore be directly linked to the policies currently pursued by the European Union in general and its leading country, Germany, in particular. The hope of policymakers is—as was initially the case in the 1930s as well—that austerity eventually will make the European economy more 'healthy' and breed the conditions for recovery. This overlooks the fact that the primary reason for the current malaise is not market imperfections that can be quickly eliminated (although financial recklessness in the lead up to 2007 clearly played a significant part), but deep-rooted differences in the capacities of different European countries to adapt to—and profit from—ongoing changes to the global economy, as we shall see in Chapter 2. Austerity will most probably not lead to improvements in this respect but rather make the situation worse by creating a hostile climate in which much needed innovative initiatives stand little chance of succeeding. A new policy stance is therefore required, and the purpose of this book is to contribute to the development of this.

1.3. Europe Facing the Climate Challenge

As in other parts of the world, Europe also faces the looming threat of climate change. The effects of greenhouse gas (GHG) emissions are slowly developing, global in character and often far from easy to identify and distinguish from natural variations. Although climate change is not a new phenomenon (it was seriously discussed in the mid-nineteenth century, for example, and the US started to systematically collect GHG data on Hawaii in 1958), climate policy discussions—and consequently research—are of more recent origin. In particular, the Intergovernmental Panel on Climate Change (IPCC)—the scientific hub around which much of climate research and policy recommendations revolves—was only established in 1988 (Bolin, 2007).

Until recently there were significant differences in opinion among scientists as to the existence of climate change, the consequences of GHG emissions, and the role of humankind in the processes taking place. Now there seems to be a growing consensus among analysts that the globe is heading towards a 3–6°C warmer Earth than a century ago, and that this global warming is to a large extent caused by GHG emissions with their origin primarily in human activities (IEA, 2011; IEA, 2013b; IPCC, 2013a, b, 2014a, b; World Bank, 2012, 2013). The concern is that, if the present emissions of GHGs in the atmosphere are not significantly reduced during the decades ahead, the consequences for human life may be dire. Large parts of the Earth may become either too dry (or too hot) to live in, or be drowned by the sea, the level of which may rise by up to one metre during this century. Moreover, all parts of the world are likely to be affected by more extreme weather conditions (IPCC, 2013a, b; World Bank, 2012, 2013).

Even though there is growing consensus among scientists on climate change, political action on how to adapt to or mitigate climate change has been lagging. The failure hitherto to reach any substantial agreements within the United Nations Framework Convention on Climate Change (UNFCCC), which started in 1992, is a prime illustration of that. There appear to be many contributing factors.

(1) Fossil fuels are the dominant source of energy in many countries. Most people have no vision of how a decent life or a functional economy can be achieved without burning fossil fuels. As discussed in Chapter 6 by Laestadius, this 'addiction to carbon' is a major factor behind the reluctance towards shifting the economy in a sustainable direction, and it is reinforced by resistance to change from dominant incumbent actors such as energy firms (see Chapter 4 by Geels and Chapter 7 by Lauber and Jacobsson).

(2) Linked to this is a reluctance among the public to engage in substantial behavioural change (including making sacrifices such as international travel), especially when the beneficiaries of such changes are remote either geographically (e.g. living in low-lying parts of Bangladesh) or temporally (i.e. future generations).

(3) Fossil fuels also provide significant employment, income, and profits, and powerful actors in affected industries or locations inevitably tend to resist change (see Chapters 4 and 7). Many regions are heavily dependent on mining of coal or lignite, and the automotive sector—still very reliant on fossil fuels—is, not the least in Europe, a strong generator of employment, both directly and indirectly.

(4) Fossil fuels are still the cheapest source of energy in many countries. Indeed, they continue to be directly or indirectly subsidized in many instances—and not only outside the OECD area (IEA, 2011, 2013b). Many governments are nervous about the political consequences of higher energy prices, and not just in poorer economies.

(5) Business and politicians alike are concerned that too restrictive policies on the use of fossil fuels in their own region or country may allow firms in countries with more lax regulations to capture market share and employment.

(6) Two hundred years of CO_2 emissions have contributed significantly to the wealth of the developed part of the world. The developing countries of today naturally want to achieve similar benefits to those reaped by the early movers and have hitherto—in the absence of cheap, sustainable alternatives—been understandably reluctant to share the burden of reducing CO_2 emissions. Although the costs of renewable energy have fallen in recent years, prices of fossil fuels have also declined significantly. So this continues to be a dilemma, as evidenced by the UNFCCC negotiations in Lima in December 2014.

Europe is in the midst of this. As Figure 1.3 shows, Europe is the third largest CO_2 emitter in the world, although on a per capita basis European emissions are not particularly high.

However, the fact that other countries emit more per capita than Europe does not mean that Europeans can sit back and wait for other countries to reduce their emissions. It is the absolute amount of greenhouse gases in the atmosphere that matters for the climate, and the world is rapidly approaching the limit for how much greenhouse gases the atmosphere can absorb if

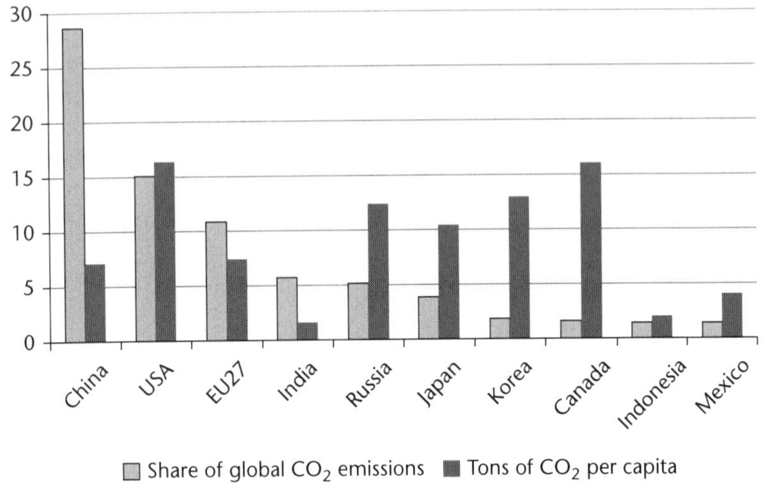

Figure 1.3. Largest global CO_2 emitters, 2012

Note: Authors' calculations based on data from the Edgar database (<http://edgar.jrc.ec.europa.eu>, accessed Sept. 2014)

damaging climate-related consequences are to be avoided (by staying within the 2°C target). According to the IPCC, in order to confine temperature rises to 2°C, global GHG emissions have to be reduced substantially by 2050, and almost completely eliminated by the end of the century (IPCC, 2014). To achieve these very demanding goals, instead of increasing the burning of fossil fuels every year, as has been the case for about 200 years, use of fossil fuels has to be reduced by approximately 3–4 per cent annually for the rest of this century (Smil, 2010). It would seem only reasonable that the developed part of the world, which has benefited from previous emissions, should shoulder the largest burden.

European politicians pride themselves on having already substantially reduced greenhouse gas emissions and hence for being on broadly the right track (European Council, 2014b). But is this correct? To explore this, Figure 1.4 traces the development of European GHG emissions from 1990 onwards.

In Figure 1.4, total EU emissions are divided into three categories: Eastern Europe, Germany (including the former GDR), and the rest of Europe. What the figure shows is that, for Europe as a whole, there was a reduction in emissions in the early 1990s, but this can be almost entirely explained by the rapid changes that took place (including the closure of inefficient plants) in the previously socialist countries in the East. For the rest of Europe, emissions were essentially stable until the outbreak of the financial crisis. The question that this raises is whether the more recent decline in emissions represents a shift towards a new, more sustainable path, or whether it is

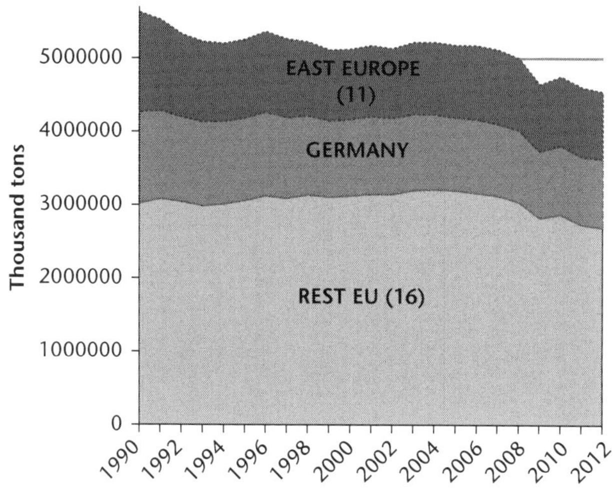

Figure 1.4. GHG emissions, EU28, 1990–2012

Note: Authors' calculations based on data from Eurostat (<http://appsso.eurostat.ec.europa.eu/ [env_air_gge]>, accessed Nov. 2014)

mainly a consequence of the crisis, and hence is likely to be reversed should the economy recover.

To investigate this, Figure 1.5 includes data on GHG emissions and growth of GDP for the European Union as a whole between 1995 and 2012. It reveals that the GHG intensity (i.e. GHG emissions per unit of output) has declined steadily, as it has in the US (Nordhaus, 2013). But until shortly before the financial crisis, this decline was not enough to reduce Europe's emissions. Moreover, as the figure shows, had growth continued at the same pace as before the crisis, emissions would have stayed roughly constant. Thus, the recent decline in emissions unfortunately does not seem to reflect a change towards a more sustainable path for the European economy, but is mainly a reflection of the continuing economic stagnation.

This raises extremely serious questions about Europe's ability to cope with these challenges. If European GHG emissions are to be reduced to a negligible level towards the end of the current century, as European policy-makers several times have agreed to, they must be reduced by approximately 3.6 per cent every year on average. Alternatively, if the GHG intensity continues to decline, as in the past, by 2.5 per cent per year, this means that Europe's GDP would have to shrink every year in order to reach the target. Hence, a revival of the economy, which is needed to reduce unemployment and increase welfare, appears to be in direct conflict with the need to combat climate change. Or to put it differently, to realize both objectives, the European economy has to be

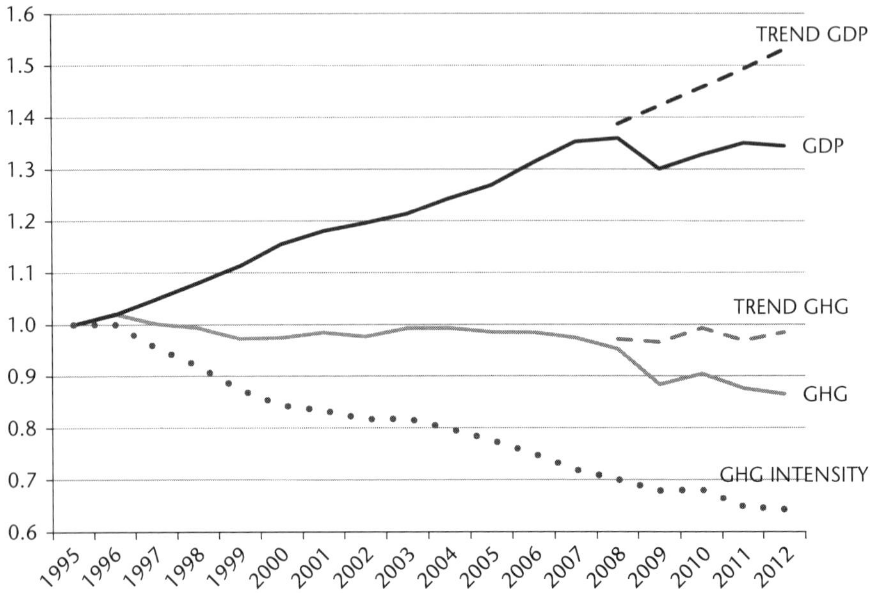

Figure 1.5. GDP and GHG emissions, EU28, 1995–2012

Note: Authors' calculations based on data from Eurostat (<http://appsso.eurostat.ec.europa.eu/ [nama_10_gdp] and [env_air_gge]>, accessed Nov. 2014). Trend-GDP is a continuation of the 1995–2007 trend for GDP, while Trend-GHG is what the emissions would have been in that case (with actual GHG intensity)

shifted to a completely new trajectory when it comes to the emission of greenhouse gases. For example, to allow for economic growth of 1.5 per cent per year for Europe as a whole, the annual decline in the GHG intensity needs to double, to around 5 per cent per year. This is a truly formidable challenge. As described by Laestadius in Chapter 6, it will require a fundamental transformation of the European economy. Activities and technologies that (directly or indirectly) emit GHGs will have to be phased out and substituted with sustainable alternatives. Sometimes this might result in new fuels being substituted for old ones (for cars, for example), but it may also involve more radical changes. Not least, significant energy savings will be necessary (IEA, 2013a, b). Creating the conditions, incentives and acceptance for this will require innovative policies.

Are there signs of such a transformation happening? After all, the policy challenge from climate change has been on the agenda for quite some time, and ambitious goals have been announced and new policy instruments created, notably the establishment a decade ago of a European market for GHG emission (the ETS—see Chapter 8 by Begg). However, the ETS has completely failed in terms of creating a price level for GHG emissions that would stimulate a transition towards increased sustainability. As is argued in Chapter 6 by

Laestadius, CO_2 prices have to increase to around €40/ton or more in order to contribute to this transformation. Yet throughout 2014 they only sporadically exceeded €7.50/ton.

In October 2014, EU leaders met to consider European policies on climate change and agreed to reduce GHG emissions by 40 per cent (compared to the 1990 level) by 2030 (European Council, 2014b). If one takes into account that half of this has been achieved already, this agreement implies an annual 1.8 per cent reduction in emissions between 2014 and 2030, about half the annual reduction needed to reach the long-term target that both the European Council and the European Parliament agreed to five years earlier. Hence, current EU leaders would seem to have ducked their responsibility, tacitly opting to leave most of the required efforts to future generations, by which time the task, according to IPCC analyses, will have become much harder (and the costs much higher).

Nevertheless, we argue that it may still be possible to stay within the 2°C target for the globe as a whole. The problem is not so much a lack of energy per se; every hour the Earth receives more energy in the form of radiation from the sun than humankind uses during a whole year (Sandén, 2008). Nor is it lack of technology—a range of technologies (solar, wind, fuel cells, systems for energy savings, etc.) are already available and becoming ever more efficient. Rather the problem is a lack of vision, creativity, and political will (Calderon and Stern, 2014). Moreover, all this has to occur rather rapidly in order to minimize the adverse consequences of climate change. To speed up that transformation process, powerful policy instruments are needed, as discussed in Chapter 9 by Mazzucato and Perez. To succeed, policy-makers have to show that climate and unemployment mitigation can be made compatible with improved welfare, albeit not a lifestyle based on the burning of fossil fuels (see Chapter 6).

For sustainable energy sources to be fully exploited, and for the associated technologies to be widely deployed and deliver on their promise, a lot of learning and—as a result—innovation has to take place. Since the outcomes of such learning and innovation processes cannot be known beforehand, and because a number of different energy sources and technologies are likely to be needed to substitute for fossil fuels, it is important to encourage experimentation to explore different trajectories, as in the case of the German *Energiewende* analysed by Lauber and Jacobsson in Chapter 7. As these two authors point out, the common criticism of *Energiewende* for supporting the development of other options besides the currently cheapest one is short-sighted. It fails to appreciate the risks (and the potential costs to the economy) of once more becoming locked into just one or a few technologies or energy sources which may not be sufficient or be available when needed (amongst other things due to the long time it takes for such technologies to mature).

1.4. The Governance Challenge Faced by Europe in Arriving at Policies for a Sustainable Future

In this section, we examine the governance challenge faced by Europe with regard to developing the necessary policies for economic recovery and confronting issues related to climate change and sustainability. Simply pumping up demand would quickly come into conflict with climate concerns and hence not be sustainable. Therefore a policy aiming at faster growth in Europe in order to increase employment and welfare must simultaneously speed up the transformation to a sustainable economy. The best way to achieve this, we argue, is to target innovation, diffusion of new technology, and, not the least, transformative investments in areas such as energy supply and distribution, increased energy efficiency, public transport, infrastructure for cars driven by electricity and fuel cells, and so on. Many of these investments in the energy sector, for example, would be necessary anyway (ECF, 2013) but undertaking them sooner rather than later (and using reduced GHG emissions as a yardstick in the selection process) may accelerate the transformation while at the same time reviving growth.

While such transformative investments are needed in all member countries, it is natural to give priority to those countries that have further to go with respect to achieving sustainability. As Figure 1.6 shows, the countries most in

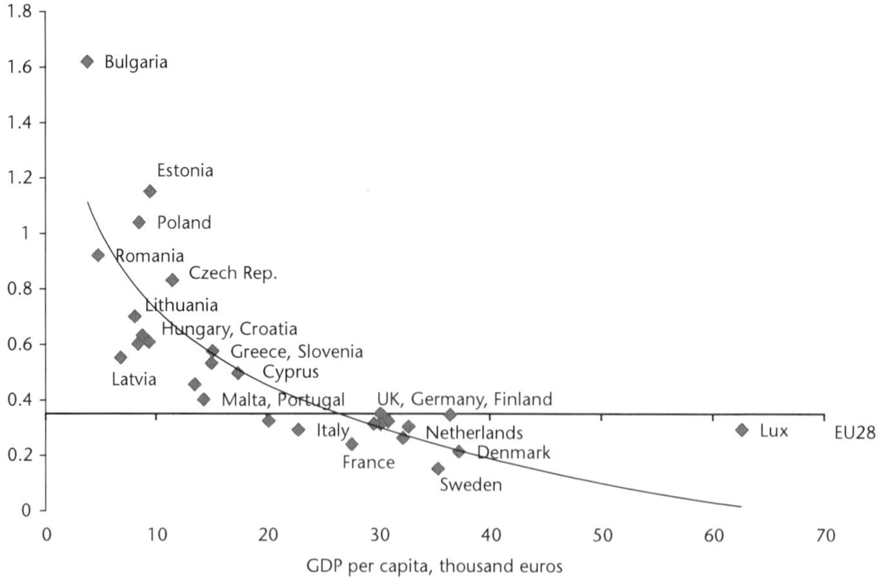

Figure 1.6. GHG intensity and GDP per capita, EU28, 2012

Note: Authors' calculations based on data from Eurostat (<http://appsso.eurostat.ec.europa.eu/ [nama_aux_gph]>, accessed Sept. 2014, and [env_air_gge], accessed Nov. 2014)

need of transforming their economies in the direction of sustainability are poorer member countries in the East and to some extent the South (which are also in general the ones most adversely affected by the current stagnation). Thus, a programme for transformative investment based on these principles would not only be good for climate change and economic growth generally, but also deliver growth where it is most needed, thereby contributing to improved social cohesion in the Union as a whole.

As several chapters in this volume point out,[4] boosting Europe's economy and its transition to a sustainable 'green' economy through transformative investments should be seen as a core element of European policy for innovation and growth. Innovation is not, as sometimes presented, primarily about creating new ideas that may (or may not) have profound effects sometime in the distant future. That is the role of research, which, although important, is not the same as innovation. Rather innovation is about trying out new, promising solutions and ways of doing things in practice, learning from the experience, and on this basis improving the technologies in question (Fagerberg, 2004). Hence, a good innovation policy is one that also promotes diffusion, use, learning, and continuous improvement. The policy approach for Europe recommended in this volume falls squarely within such a framework.

Yet the challenges in achieving all this are substantial. Indeed, the governance challenges have become all the more daunting in recent years for a number of reasons, some of a general nature, others more specific to Europe. In what follows, we consider the implications of these in more detail. The first has to do with the increasingly global nature of the problems confronting governments, requiring internationally coordinated, multilateral efforts to overcome them. Climate change and economic development are prime examples of that. The failure so far to come up with a comprehensive international agreement on how to deal with climate change is, of course, very disappointing. Nevertheless, as Laestadius points out in Chapter 6, there may be possibilities for international cooperation affecting climate change even in the absence of an effective global treaty, such as alliances of like-minded countries (with Europe in the lead) pioneering new solutions and encouraging others to follow. There may also be possibilities for creating new forms of cooperation based on mutually recognized interdependencies, such as those between China and Europe in renewables, as highlighted by Schmitz and Lema in Chapter 5.

Second, not only are the political problems to be confronted today larger-scale, but they are also more likely to cross-cut organizational boundaries, particularly within governments (Bauer et al., 2012, 281–2). Challenges such as climate change or unemployment do not fit easily within a single ministry

[4] See in particular Ch. 6 by Laestadius and Ch. 9 by Mazzucato and Perez.

or department. Moreover, the problems tend to interact in an increasingly complex manner. For example, as already noted (see also Chapter 6 by Laestadius), policies to lower unemployment tend to lead to greater CO_2 emissions, unless carefully integrated with climate change policies. Energy policy, for instance, must (as Geels points out in Chapter 4) give careful consideration to a range of issues, including security, for example. Effective policies for transforming the economy towards sustainability will most probably require the development of new forms of governance, characterized by a holistic perspective and by close coordination between different parts of government.

Third, increasing globalization and the growing prominence of civil society have weakened the influence and power of the nation state (Holton, 1998; Pahl-Wostl, 2009). Over time, particularly in Europe, there has been a shift from dominance by national-level governments to multi-level governance (Hooghe and Marks, 2001)—local, regional, national, EU, and in some cases global. However, although this clearly adds to the complexity of governance (e.g. Bache, 2012), it also means that local and regional actors have much more space than before for pursuing their own strategies and forming alliances with others in similar situations, for instance, with respect to mitigating climate change in the case of cities (as Geels emphasizes in Chapter 4).

A fourth issue regards the increasing involvement of non-government actors, not least in Europe. The challenge associated with climate change illustrates the growing range of non-government actors that need to be involved in addressing the problem in order to ensure their commitment to the resulting policies and their implementation (Biermann, 2007, 332–3; Biermann and Pattberg, 2008, 280–2; Biermann and Gupta, 2011, 1856; Bauer et al., 2012, 282–3). However, while making governance more complex, the involvement of non-governmental actors may introduce a much needed new dynamic in policy-making, as evidenced, for example, by the Green movement in Germany with respect to *Energiewende* (see Chapter 7 by Lauber and Jacobsson).

Fifth, there is a heightened sensitivity to risk and uncertainty, again a trend that has arguably been more pronounced in Europe than elsewhere. Technological advance and environmental change are areas beset by persistent uncertainty regarding their causes and their interactions, their impact and possible response options (Biermann, 2007, 329). Risk and uncertainty naturally lead to advice advocating caution (with respect to nuclear energy, for example). However, the fundamentally uncertain nature of technological advance also means that policies for transformation should, as pointed out above, place an emphasis on pursuing a broad portfolio of different energy technologies and not getting locked into a specific path of development that at one point may appear more cost-effective or promising.

A sixth factor adding to the governance challenges facing the EU is the growing number of Member States and the fact that they are at various stages of economic development and with different economic, industrial and other strengths and weaknesses (see Chapter 2 by Fagerberg and Verspagen, and Chapter 3 by Landesmann). In the beginning, the European Economic Community consisted of just six members, each at a broadly similar level of development. Now, with twenty-eight Member States and a number of other 'neighbours' who are aspiring members or otherwise subject to EU policy influence, those countries are quite different in terms of economic, industrial, and institutional characteristics. It is no longer clear that policies based on the philosophy of 'one size fits all' are appropriate for the EU. On the contrary, as Fagerberg and Verspagen suggest in Chapter 2, policies that do not take Europe's heterogeneous character properly into account may well end up further aggravating the differences between countries rather than improving social cohesion. This applies, for example, to the (necessary) transformation away from the cheap coal that some poorer Member States depend on.

Seventh, and again a factor specific to the governance challenges faced by the EU, is the fact that the scale of resources at the disposal of the EU is in most cases very limited compared with those allocated by national governments, as Begg stresses in Chapter 8. For example, in the case of R&D and innovation, national government spending by Member States is typically an order of magnitude greater than that by the European Commission (as is spending by the private sector). The EU must, therefore, attempt to carefully 'leverage' its efforts, as Begg argues, using its political authority to try to influence and coordinate national governments in a coherent and effective manner.

Lastly, the governance challenge and the task of arriving at a suitable policy response is made more difficult by growing concerns about a democratic deficit (Harlow, 2002; Follesdal and Hix, 2006) and a declining trust in (and diminishing popular support for) European institutions (see Chapter 8 by Begg for evidence on this). The spreading disillusion with the EU is to a large extent a consequence of the troubled economic circumstances and the apparent inability of EU policy-makers to do much to alleviate this. Hence, the failure of EU politicians in dealing effectively with the challenges that Europe faces is now coming back to haunt the entire European project. This clearly underscores the need for a new policy stance.

1.5. Where Do We Go from Here?

The central argument in this book is that, for Europe and elsewhere, the climate challenge and the economic challenge are intimately interrelated.[5]

[5] The interrelated nature of these challenges, obvious at it may seem, has apparently not yet penetrated EU policy thinking to any great extent. For example, in the conclusions from the

Without growth, greenhouse gas emissions may continue to decline, although almost certainly not at the pace necessary for their elimination by the end of the present century. But employment would undoubtedly continue to suffer. A return to the type of growth that prevailed before the 2007–8 financial crisis, on the other hand, might be good for employment but certainly not for the climate, as emissions would continue at unsustainably high levels.

Is there a way out of this dilemma? For a start, economic growth does not necessarily mean doing more of the same. It can also mean getting more out of a given level of resources by doing things in a smarter way than before—that is, by innovation. Indeed, since the time of Robert Solow, who in 1970 won the Nobel Prize in economics for his work over the previous fifteen years on economic growth, economists have consistently argued that innovation is the major source of economic growth in the long run.[6] Some are sceptical with regard to the potential of future innovation. But are there really reasons to believe that humankind is less capable of coming up with smart solutions than, say, fifty or hundred years ago? We believe the answer to this question is a clear 'No'. Smart solutions, or innovations, typically mean combining different sources of information, knowledge, and other resources in a novel way,[7] for which education is a great advantage (if not a must). Never have levels of education been as high as they are today. The same holds for the share of global economic resources used in searching for new and better ways of doing things.[8] What we are arguing for, therefore, is more use of 'brain power' and of sustainable forms of energy, substituting for the 'fossil power' that has dominated economic and technological development for two centuries.

Nevertheless, as Robert Gordon (2012) has argued, even if technological progress continues unfettered over the remaining part of the twenty-first century, there may well be various countervailing factors—what he calls 'headwinds'—resulting in substantially slower growth in income (or GDP per capita). He points specifically to demography (and an ageing population), reduced scope for increasing the qualifications of the labour force (through investments in education), globalization, increasing inequality (which he identifies as arguably the most important 'headwind'), climate change, and the large public and private debts that need to be reduced. According to

European Council meeting of 23 and 24 Oct. (European Council, 2014a), climate change and stagnation are both highlighted, but there is no mention of their interrelated nature.

[6] See Fagerberg (1994) and Verspagen (2004) for more detailed accounts.

[7] Schumpeter, who is generally regarded as the founding father of the field of innovation studies, frequently used the term 'new combinations' instead of 'innovation' in his early work.

[8] E.g. between 1996 and 2012, the share of GDP devoted to R&D increased by one-third globally, and in China R&D expenditure as a percentage of GDP almost tripled over the same period to around 2%, i.e. about the same level as the EU average (based on data from UNESCO <http://data.uis.unesco.org>, accessed Dec. 2014).

Gordon, the joint effect of these 'headwinds' may well be to restrict future growth of GDP per capita in the United States to almost zero.

However, what (perhaps) holds for the US does not necessarily hold for the world as a whole. For example, as we have pointed out, countries at a lower level of development can gain a lot by exploiting technologies already in place in technologically more advanced nations. Moreover, an ageing population is clearly not a problem in Africa, home to some of the fastest-growing countries in recent years. Furthermore, as Gordon points out, exploiting the potential for immigration—as the USA has done so successfully in the past—may go a long way to mitigate the potentially negative effects on economic growth of an ageing population, in the US as well as elsewhere. Similarly, average education levels are still low in large parts of the world, so continuing investment in education should continue to bring large pay-offs. Globalization is not just a challenge but also an opportunity. Indeed, the rapid economic growth witnessed in large parts of the developing world in recent years has much to do with exploiting the opportunities offered by globalization. Inequality may well be a problem, but again there are substantial differences between countries in this respect, with inequality being much higher in the US than in Europe (Cingano, 2014). Moreover, this is not an exogenous factor but something politics can influence. Nevertheless, recent research on inequality and economic growth (Cingano, 2014; Piketty, 2014) points to an important lesson. Sustainable growth, to fulfil its promises, cannot just be about decreasing GHG emissions; it also needs to focus on achieving an equitable distribution of resources and benefits consistent with fostering social cohesion.

How do these factors influence growth prospects in the European case? First, as already pointed out, GDP per capita in Europe is still just two-thirds of that in the US, so there should be plenty of scope for catching up and hence for faster growth than in the US. Moreover, the differences in levels of economic development are much larger within Europe than between US states (Fagerberg et al., 2014), which means that there is even greater scope for growth in poorer EU Member States, particularly in Eastern Europe. Although the issue of declining and ageing populations has not been central in this volume, it clearly represents a challenge, with only a handful of European countries currently having fertility rates sufficient to ensure a constant population. This challenge is especially evident in Southern Europe (Portugal, Spain, and Greece), in other words, the very same countries that have suffered most in the aftermath of the financial crisis, as well as in some Eastern European countries (with Poland being the most important case).[9] Yet, as with several other challenges, this is something policy can influence, for example by creating

[9] See <http://ec.europa.eu/eurostat/statistics-explained/index.php/Fertility_statistics> (accessed Dec. 2014).

favourable conditions for families with children, and allowing for more integration of young, entrepreneurial job-seekers from countries outside the EU, of which there appears to be an abundant supply. The growing anti-immigration sentiments in several European countries are both backward-looking and counter-productive; problems with integration notwithstanding, immigration is conducive to Europe's economic future, not a threat. Moreover, it may be argued that in the short term an ageing population is not a fundamental problem as long as 25–50 per cent of the young population is unemployed.

Furthermore, while education levels are high in large parts of Europe, particularly in the North, average education levels in the South of Europe continue to lag substantially behind other European countries. Investing in education there, including the education of the current adult population who will continue to dominate the labour force for some time to come, may prove to be an essential element in a long-term policy for reviving the depressed economies in this part of Europe, not least in terms of leaving them better equipped to exploit opportunities opened up by globalization and European economic integration, opportunities which—as is pointed out in Chapter 2—they have largely failed to address so far. Finally, while the distribution of income generally is more equitable in Europe than elsewhere, Southern Europe again represents an exception.[10] Thus, a policy aimed at reviving these economies needs to pay close attention to the distributional aspects of economic development.

At the end of 2012 most EU countries were still producing less per inhabitant than four years earlier. Why has Europe been so slow in recovering from economic depression? In our view this points to important weaknesses in the construction of the EU. The major problem has been the creation of a monetary union for countries with very different characteristics and economic dynamics, and without any effective policy coordination and the necessary mechanisms for mitigating problems that may occur either as a result of the diversity of the participating countries or due to external shocks. Hence, the entire Union was singularly ill-prepared for stormy weather, something that became all too apparent when the financial crisis struck in 2007–8. As Begg points out in Chapter 8 of this volume, the economic levers available to the EU are very limited. Worse still, the adoption of the common currency prevented the most crisis-ridden countries in Southern Europe from dealing as effectively with that crisis as they might otherwise have done. Instead, their governments were forced by the governing bodies of the EU and IMF to practise stringent austerity, with economic decline, higher unemployment

[10] See <www.oecd.org/social/income-distribution-database.htm> (accessed Dec. 2014).

(especially among the young), and rising inequality the almost inevitable consequences. It would seem that the European Union has dropped all aspirations of becoming 'the most competitive and dynamic knowledge-based economy in the world' and instead, not least under the influence of its largest and most powerful member country, Germany, turned more into a cartel for stagnation.

The German economic model, exploiting demand abroad rather than generating it at home, is based on the country's industrial strength and competitiveness. However, not all countries in the world can have an export surplus. Hence, what may work for a single country does not necessarily work for the global economy, and arguably not for an entity as large and heterogeneous as Europe, either. A radically different policy stance is required if Europe is to escape the current stagnation and its extensive social and economic consequences. Such a policy stance has to take into account the fact that the economies of Europe are very different, so there is no point in just mimicking the same policy, whether patterned on German experience or that of some other country, everywhere. One size does *not* fit all. Rather, what needs to be done is to identify the factors that prevent the various economies from growing and then addressing these factors head on. For example, the characteristics of Southern Europe, with lower levels of education, weaker industrial dynamics, and higher levels of inequality, require specific policies targeting these features.

Recently, the governing bodies of the EU, in particular the European Commission and the European Central Bank, have started to devote more attention to the need to get Europe out of the present slump by increasing investments (European Commission, 2014) and adopting a more expansionary monetary policy. Although these initiatives are surely welcome, they will do little to stimulate the much needed European transformation towards sustainable growth, as discussed in Chapter 6 in this volume.[11] The crucial question for policy, therefore, is how to get the European economy out of its present depressed state and simultaneously mobilize the European population in a collective innovation journey in the direction of a sustainable economy and society.

How innovation can best be stimulated, and what its social and economic effects are, has been the focus for researchers within the field of innovation studies for half a century or more.[12] This research depicts innovation as a highly uncertain process responding to opportunities as they arise, be they in

[11] E.g. in his recent presentation of the EU investment plan, Vice-President Jyrki Katainen did not once mention climate change or sustainability (see 'Vice-President Jyrki Katainen explains the EU investment plan', <http://ec.europa.eu/priorities/jobs-growth-investment/plan/index_en.htm>, accessed Jan. 2015).

[12] See Fagerberg et al. (2004) and Martin (2012) for an overview.

the form of new knowledge (which may make new products, services, or ways of doing things possible) or related to emerging problems, needs, and challenges (which require new solutions to be found). Innovation policy, therefore, is partly about supporting the development and spread of new knowledge that private sector firms would not develop with their own money, but which nevertheless has the potential to spur subsequent innovation and economic growth. The creation of the internet (by the US Government) is a good example (see Chapter 9 by Mazzucato and Perez). However, creating demand for innovative new solutions may be equally or even more crucial because, without demand, innovative firms get nowhere (and are very likely to abandon their innovation efforts). Indeed, if demand is lacking, traditional support schemes such as subsidies or cheap loans may prove largely ineffective. Hence, governments can exert great influence on innovation activity by emphasizing the challenges that need to be addressed, or the problems that need to be resolved, provided this is matched by a credible commitment to generating sufficient demand so that the eventual solutions are adequately rewarded. This type of innovation policy, which provides a sense of direction to the collective innovation journey and rallies potential contributors behind it, would be relevant for a wide range of activities essential for the transition to a sustainable economy, such as energy production, distribution, and use, and also transport and construction. In order to be effective, such a policy must place different policy arenas (energy, transport, regional development, research, innovation, etc.) under a common umbrella. Hence, sustainable growth requires more than technological innovation; new forms of governance are also required.

Europe is only the third largest greenhouse gas emitter worldwide. Given the global character of the problem, and the many actors involved all round the world at different levels who may have a say in what happens, the ability to influence actors in other countries becomes of central importance. One key way to achieve this, and one for which Europe seems eminently well placed, would be to lead by example, providing solutions for how the climate challenge can be effectively dealt with. Taking the lead may, of course, incur significant costs, which Europe as one of the richest parts of the world should be able to afford. The cost structure is far from obvious, however: doing nothing will, for sure, have an impact in the years ahead in many areas of life. Conversely, Europe taking the lead in this challenge may, as discussed in Chapter 6, not only attract followers but also result in substantial benefits in the form of developing industrial competitiveness, exports, and new jobs. Following this path would mean turning Europe into a veritable laboratory for sustainable growth, environmentally as well as socially. This is the direction we advocate that European policy-makers should take.

1.6. The Structure and Argument of the Book

Given the nature and scale of the economic, climate, and governance challenges now facing Europe, what is the appropriate policy response? In the chapters that follow, we examine the extent to which Europe has or has not succeeded in handling the various challenges with appropriate policies, and consider the roles of different types of policies and actors.

The Economic Challenge

Chapters 2 and 3 reveal how current austerity policies are not socially sustainable, having resulted in high levels of unemployment and little or no growth and consequently a 'lost decade' for Europe. It is shown that the problem is not so much the lack of a well-functioning financial sector (although that has undoubtedly contributed to the problems) as a lack of competitiveness among countries at the periphery of Europe and a lack of demand.

Chapter 2 sets out clearly how the European economy is currently in a slump, the worst since the 1930s. Although this is often seen as a consequence of the financial crisis that hit the capitalist world in 2007–8, Jan Fagerberg and Bart Verspagen reveal that many of the problems that Europe faces today have longer term roots to do with the fact that Europe consists of countries with quite different dynamics and capacities for adapting to changes in the global (and European) economic environment. They start by comparing Europe's growth performance with that of other parts of the world, and then consider some popular but seemingly erroneous explanations of the present crisis. Subsequently, they identify three European 'archetypes', characterized by different levels of adaptability and performance, i.e. the North, the South, and the East. The chapter explores the consequences of globalization and European economic integration for the economic performance of these country groups. The effects have been quite asymmetric, with the Southern countries in particular having benefited little if at all. Finally, the authors summarize the lessons from their analysis and consider the implications for policy. They argue that what is needed is a European growth policy, properly adapted to the varying characteristics and capabilities of different European countries, a policy that places the welfare of the European population as a whole at the centre.

In Chapter 3, Michael Landesmann shows how the European Union had built up a track-record for gradually reducing the economic differences across Member States in the decades leading up to the recent economic crisis. This prompted talk of a successful European integration model that included not only existing EU member economies but also a broader range of countries with strong connections to the EU. The chapter discusses the foundations of the integration model and the reasons why it was seriously derailed in the period prior to the crisis. A significant range of lower and middle income economies experienced the build-up of unsustainable external imbalances, an accelerating rise in their debts (mostly in the private sector), and major distortions in their economic structures to the detriment of the tradable sector. Several economies have become dependent on a small range of commodity exports and suffer from a lack of diversification. The author argues that the policy framework associated with EU integration has to be reshaped in order to avoid a persistent divide between the richer countries and the rest of the Union and its neighbourhood.

The Climate Challenge

Chapters 4, 5, and 6 are concerned with the climate challenge, how Europe has dealt with it, what is required to transform the economy in a sustainable direction, and the role of different actors in this context. It is shown that Europe's current path is not consistent with its stated aim of almost completely eliminating the emission of greenhouse gases by the end of the present century. We also see that European climate policies have been well-intended but often ineffective, the emission trading system (ETS) being a case in point, and that a new policy stance is required. The problem, it is argued, is not primarily a lack of knowledge but a lack of political will and concerted action.

In Chapter 4, Frank Geels discusses the arduous transition to low-carbon energy supply, with a focus on electricity generation. The chapter first makes an assessment of global and European developments, addressing not only the expansion of renewable electricity capacity (wind, solar-PV, biomass) and city-based initiatives, but also the ongoing expansion of coal-fired power generation, the shale gas revolution, and conflicting trends in nuclear energy (phasing out in some countries, expansion in others). Geels argues that the resilience of existing regimes is preventing a diffusion of renewables that is sufficiently rapid to limit climate change to 2°C. Subsequently, the chapter focuses on the UK to further analyse the resilience of existing electricity generation regimes, particularly the strategies of utilities and the national government, which are shown to be firmly committed to nuclear, gas, and clean coal (with the promise of carbon capture and storage, CCS). The analysis shows that incumbent actors both adapt to climate change pressures (through innovation) and attempt to resist major reorientation (through political strategies). The chapter ends by discussing wider challenges. It is noted that the world's proven fossil-fuel reserves far exceed the amount of carbon that can be burned while staying within the 2°C target. Hence, Geels argues, low-carbon transitions are not just about stimulating 'green' alternatives, but also about actively managing the decline of fossil-fuel-based systems, and that much bolder policies will therefore be required.

Chapter 5 by Hubert Schmitz and Rasmus Lema addresses the challenge from economic globalization for the green transformation in Europe. It does so by analysing the economic rise of China as a key feature of globalization in the wind power industry. The effects on the European wind power industry are shown to be different from those in the solar photovoltaic industry. While the latter declined due to China's exports, the European wind power industry continues to be strong in terms of innovation and production. Although also challenged by China, the relationships between European and Chinese firms are characterized by both intense competition and close collaboration. Schmitz and Lema show how the global financial crisis influenced the Chinese challenge to Europe, and reflect on how the new division of labour affects the chances of mitigating climate change. While Europe continues to be the driving force in innovation, China is playing a key role in bringing down the costs of renewable energy. Hence, the increasing role of China in renewable energy technologies should not (only) be seen as a threat, but also as an opportunity for European industry and an essential element in accelerating the green transformation.

In Chapter 6, Staffan Laestadius analyses the conditions for a transition path away from the current European crisis characterized on the one hand by persistent high unemployment, and on the other hand by GHG emissions several times greater than can be allowed for EU countries if the 2°C target for the global temperature increase is to have a reasonable chance of being met. For 200 years, economic growth has been strongly correlated with growing energy demand in general and with increased use of fossil fuels in particular. Thus, burning of fossil fuels not only contributes to much of existing employment; it is also a core component in modern welfare. The problem can hence be formulated in terms of steering 'between Scylla and Charybdis': we need to 'pump up demand' to increase employment—but if we do, this will increase GHG emissions, thereby pouring fuel on the fire. The chapter is focused on the potential for achieving a structural transformation towards sectors in the economy that are either green (i.e. non-GHG-emitting) or blue (providing the infrastructure for green activities), while at the same time bringing about the Schumpeterian 'creative destruction' of the black sectors (currently by far the largest part of the economy, and the part that is responsible for GHG emissions). The discussion concentrates on the important energy and transport sectors, the transformation of which may arguably have a major knock-on impact on others.

Policies for a Sustainable Europe

Chapters 7, 8, and 9 deal with the question of what policies for a sustainable Europe might look like. It is argued that new forms of innovation policy aimed at inclusive and sustainable growth will be required, and that government needs to play a central role, not the least by providing direction for innovation and entrepreneurship. One example of such a policy—the German *Energiewende*—is considered in some detail. The potential role of the European Union in the necessary transformation is also explored.

Chapter 7 by Volkmar Lauber and Staffan Jacobsson focuses on the German policy for energy transition, or *Energiewende*. It was designed as a long-term strategy to secure the necessary resources and support for a transition to sustainability in energy supply (through denuclearization and decarbonization). It was also a response to the challenge of globalization in that it aimed to improve Germany's competitive position by reducing fossil-fuel imports and stimulating the development of new industries. Hence, *Energiewende* was designed to meet the triple challenge in the field of energy and, as such, may provide insights into what such transformation processes entail and what policy tools are needed for managing such a radical and long-term enterprise. During its first decade, *Energiewende* was successful in stimulating the deployment of a range of new technologies, the formation of innovative capital goods industries, and securing public support. Since 2010, however, there has been growing disagreement about *Energiewende* among political actors, electricity incumbents, and energy-intensive industries, which has led to a reassessment by political elites and a slowdown of *Energiewende*, despite continuing support from the public. Lessons are drawn from the German experience of managing such a huge project, a project demanding precisely the kind of long-term policy, staying power (to overcome the inevitable opposition in certain quarters), and public support that the transformation to a sustainable Europe will require.

Chapter 8 by Iain Begg examines whether previous efforts by the EU to confront grand challenges such as climate change and competitiveness have been a help or a hindrance. The chapter distinguishes between policies implemented directly at the EU level of governance and policy orientations that emerge within the EU as a result of complex interactions among Member States with their different priorities and preferences. Begg examines the range of ways in which the EU affects and influences Member States, and offers an assessment of some of the principal policy strategies of recent years. He highlights how there can be unrealistic expectations of what the EU is constitutionally and financially able to do, identifying certain tensions that have arisen in the EU's handling of the crises of recent years. In assessing how effective EU policies relating to the main themes of this book have been, the chapter finds that the EU has a distinctly patchy record. It argues that while the diagnoses emanating from the EU level have generally been well-founded, too many of the ensuing policy interventions have lacked the financial or political clout to make a real difference.

In Chapter 9, Mariana Mazzucato and Carlota Perez argue that the advanced world is facing a crucial moment of transition. A successful outcome requires bringing innovation to the centre of government thinking and action. In order to do this, we must apply our knowledge of how innovation occurs and how to repair what has gone wrong. The chapter looks first at the role that innovation has played as the driver of economic growth, and at its relationship with finance. Arguing that the challenge today is not to 'fix' finance while leaving the economy sick, but rather to change the way that the real economy works, the authors then identify the solution: a policy direction that is smart, inclusive, and takes advantage of 'green' as the next big technological and market opportunity. They explain why the role of the state is key to ensuring that such opportunities are taken, and highlight the importance of direct public investment for promoting the creation of public goods and for encouraging courageous risk-taking in research and innovation in both the public and private sectors. Paying particular attention to Europe, they then examine the potential of such innovation-oriented policies to promote inclusive growth. Mazzucato and Perez conclude with a number of concrete steps that could be taken, at both the national and EU levels, to create the 'smart governance' necessary to implement such a direction.

1.7. Summary

Europe is confronted by an intimidating triple challenge—economic stagnation, climate change, and a governance crisis. This book demonstrates how these three challenges are closely interrelated. In particular, a return to economic growth cannot come at the expense of greater risk of irreversible climate change. Instead, what is required is a reconceptualization of what is intended by 'economic development' and a fundamental transformation of the economy to a new 'green' trajectory based on the rapidly diminishing emission of greenhouse gases. This entails a much greater emphasis on innovation in all its forms (not just technological). Innovation policy must be placed at the very heart of industrial policy and indeed of economic policy more broadly. Other parts of the world are also facing varying forms of the triple challenge, although the governance challenge may not be quite as severe as for Europe and the EU. However, Europe is uniquely placed to take the lead in addressing the triple challenge. While this may well entail certain costs in the short term, it will undoubtedly bring considerable benefits in the longer term in the form of strengthened industrial competitiveness, enhanced exports, and new jobs. It should also encourage countries in other parts of the world to follow Europe's lead in this transformation process, thereby ensuring that climate change is kept within manageable bounds. Addressing the triple challenge would thus provide Europe and its citizens with a new sense of purpose, revitalizing the EU and 'the European project' over the decades to come.

References

Abramovitz, M. (1986). 'Catching up, Forging ahead, and Falling behind', *Journal of Economic History*, 46: 386–406.

Abramovitz, M. (1994a). 'The Origins of the Post-War Catch-up and Convergence Boom', in J. Fagerberg, B. Verspagen, and N. von Tunzelmann (eds), *The Dynamics of Technology, Trade and Growth*, Aldershot: Edward Elgar, 21–52.

Abramovitz, M. (1994b). 'Catch-up and Convergence in the Post-War Growth Boom and After', in W. J. Baumol, R. R. Nelson, and E. N. Wolf (eds), *Convergence of Productivity: Cross-National Studies and Historical Evidence*, Oxford: Oxford University Press, 86–125.

Bache, I. (2012). 'Multi-Level Governance in the European Union', in D. Levi-Faur (ed.), *The Oxford Handbook of Governance*, Oxford: Oxford University Press, 628–41.

Bauer, A., Feichtinger, J. and Steurer, R. (2012). 'The Governance of Climate Change Adaptation in 10 OECD Countries: Challenges and Approaches', *Journal of Environmental Policy and Planning*, 14: 279–304.

Biermann, F. (2007). 'Earth System Governance as a Crosscutting Theme of Global Change Research', *Global Environmental Change*, 17: 326–37.

Biermann, F., and Gupta, A. (2011). 'Accountability and Legitimacy in Earth System Governance: A Research Framework', *Ecological Economics*, 70: 1856–64.

Biermann, F., and Pattberg, P. (2008). 'Global Environmental Governance: Taking Stock, Moving Forward', *Annual Review of Environmental Resources*, 33: 277–94.

Bolin, B. (2007). *A History of the Science and Politics of Climate Change*, Cambridge: Cambridge University Press.

Calderon, F., and Stern, N. (2014). *Better Growth, Better Climate: The New Climate Economy Report*, Global Commission on the Economy and Climate, Washington, DC: World Resources Institute.

Cingano, F. (2014). *Trends in Income Inequality and its Impact on Economic Growth*, OECD Social, Employment and Migration Working Papers, 163, Paris: OECD Publishing, <http://dx.doi.org/10.1787/5jxrjncwxv6j-en>, accessed Jan. 2015.

ECF (2013). *From Roadmaps to Reality: A Framework for Power Sector Decarbonisation in Europe*, Brussels: European Climate Foundation.

European Commission (2014). *An Investment Plan for Europe*, Brussels: European Commission, <http://ec.europa.eu/priorities/jobs-growth-investment/plan/index_en.htm>, accessed Jan. 2015.

European Council (2014a). *Conclusions from the European Council 23 and 24 October 2014*, EUCO 169/14, Brussels: European Council, <http://register.consilium.europa.eu/doc/srv?l=EN&f=ST%20169%202014%20INIT>, accessed Jan. 2015.

European Council (2014b). *2030 Climate and Energy Policy Framework, Conclusions*, EUCO 169/14, CO EUR 13 CONCL 5, Brussels 23–24 Oct. 2014.

Eurostat (2014). *Greenhouse Gas Emissions*, <ec.europe.eu/Eurostat/data/database/env_air_gge>, accessed Jan 2015.

Fagerberg, J. (1994). 'Technology and International Differences in Growth Rates', *Journal of Economic Literature*, 32: 1147–75.

Fagerberg, J. (2004). 'Innovation: A Guide to the Literature', in J. Fagerberg, D. Mowery, and R. Nelson (eds), *The Oxford Handbook of Innovation*, Oxford: Oxford University Press, 1–26.

Fagerberg, J., Mowery, D., and Nelson, R. (eds) (2004). *The Oxford Handbook of Innovation*, Oxford: Oxford University Press.

Fagerberg, J., Feldman, M., and Srholec, M. (2014). 'Technological Dynamics and Social Capability: US and Europe', *Journal of Economic Geography*, 14(2): 313–37.

Follesdal, A., and Hix, C. (2006). 'Why there is a Democratic Deficit in the EU: A Response to Majone and Moravcsik', *Journal of Common Market Studies*, 44: 533–62.

Gerschenkron, A. (1962). *Economic Backwardness in Historical Perspective*, Cambridge, Mass.: Belknap Press.

Gordon, R. (2012). *Is US Economic Growth Over? Faltering Innovation Confronts the Six Headwinds*, Working Paper, 18315. Cambridge, Mass.: NBER.

Harlow, C. (2002). *Accountability in the European Union*, Oxford: Oxford University Press.

Holton, R. J. (1998). *Globalization and the Nation-State*, Basingstoke: Palgrave.

Hooghe, L., and Marks, G. (2001). *Multi-Level Governance and European Integration*, Lanham, Md.: Rowman & Littlefield.

IEA (2011). *World Energy Outlook 2011*, Paris: OECD/IEA.

IEA (2013a). *Redrawing the Energy Climate Map*, Paris: OECD/IEA.

IEA (2013b). *World Energy Outlook 2013*, Paris: OECD/IEA.

IEA (2014a). *Key World Energy Statistics 2014*, Paris: OECD/IEA.

IEA (2014b). *World Energy Outlook 2014*, Paris: OECD/IEA.

IPCC (2013a). *Climate Change 2013: Mitigation of Climate Change*, Cambridge: Cambridge University Press.

IPCC (2013b). *Climate Change 2013: The Physical Science Basis*, Cambridge: Cambridge University Press.

IPCC (2014). *Climate Change 2014: Mitigation of Climate Change*, WGIII contribution to AR5, <www.mitigation2014.org>, accessed Jan. 2015.

Martin, B. R. (2012). 'The Evolution of Science Policy and Innovation Studies', *Research Policy*, 41: 1219–39.

Nordhaus, W. (2013). *The Climate Casino: Risk, Uncertainty and Economics for a Warming World*, New Haven and London: Yale University Press.

Pahl-Wostl, C. (2009). 'A Conceptual Framework for Analysing Adaptive Capacity and Multi-Level Learning Processes in Resource Governance Regimes', *Global Environmental Change*, 19: 354–65.

Piketty, T. (2014). *Capital in the Twenty-First Century*. Cambridge, Mass.: Belknap Press.

Sandén, B. A. (2008). 'Solar Solution: The Next Industrial Revolution', *Materials Today*, 11(12): 22–4.

Smil, V. (2010). *Energy Transitions: History, Requirements Prospects*, Santa Barbara, Calif.: Praeger.

Verspagen, B. (2004). 'Innovation and Economic Growth', in J. Fagerberg, D. C. Mowery, and R. R. Nelson (eds), *The Oxford Handbook of Innovation*, Oxford: Oxford University Press, 487–513.

World Bank (2012). *Turn Down the Heat: Why a 4°C Warmer World Must be Avoided*, Washington, DC: World Bank.

World Bank (2013). *Turn Down the Heat: Climate Extremes, Regional Impact, and the Case for Resilience*, Washington, DC: World Bank.

Part I

Europe in Crisis

2

One Europe or Several? Causes and Consequences of the European Stagnation

Jan Fagerberg and Bart Verspagen

2.1. Introduction

The European economy is currently in a slump, the worst since the 1930s, with high unemployment and deteriorating welfare conditions for exposed segments of the population in several European countries, especially in the Southern part of the continent. Although this is often seen as a consequence of the financial crisis that hit the capitalist world in 2007–8, this is only part of the story. In this chapter we aim to show the following:

1. Many of the problems that Europe faces today have long-term roots and were apparent well before the financial crisis struck. Hence, the focus on this chapter will mainly be on the period leading up to the crisis.[1]
2. Far from being a homogeneous set of countries, Europe consists of countries with quite different dynamics and capacities for adapting to changes in the global (and European) economic environment.
3. As a consequence of this heterogeneity, the 'one size fits all' policies that have characterized European economic integration in the last few decades have had quite different effects in different parts of Europe. While some countries benefited considerably, others have not.
4. The current austerity measures and 'belt-tightening' in large parts of Europe are not helpful in reviving economic growth and reducing unemployment, and need to be replaced by a Europe-wide growth policy, taking Europe's heterogeneous character and the challenges it is facing fully into account.

[1] The developments during the crisis are analysed in more detail in Ch. 3 by Landesmann.

We develop these arguments in several stages. In section 2.2 we place the European growth and stagnation in its global context. During the last few decades the global economy has gone through a dramatic transformation, most importantly perhaps as a result of industrialization in countries such as China and India, and their subsequent increasing presence in global markets. This process affected European countries in different ways depending on their capacities to adapt to these changes. Similarly, the dissolution of the Soviet empire and the gradual inclusion of the previously socialist economies in Eastern Europe in the global (and European) capitalist economy have also affected European countries in different ways. In particular, there is a group of countries in Southern Europe that both in the decades leading up to the crisis, and during the crisis, underperformed compared to other countries at similar levels of development, while an opposite pattern may be observed in the Northern part of the continent. There is also a distinct Eastern growth pattern—quite different from the other two—associated with the previously socialist economies in the East of Europe.

Using these three categories, the North, the East, and the South of Europe, we show in the sections that follow that much—albeit not all—of the heterogeneity in economic development across Europe can be summarized in this way. Section 2.3 examines the widespread—though as we will show largely erroneous—idea that the problems encountered by Southern Europe have to do with a deviant macroeconomic stance characterized by excessive growth of wages and/or consumption, a so-called 'spending-spree'. The analysis shows that the relationships between the relevant macroeconomic variables did not differ much between Southern Europe and the rest of Europe or other parts of the globe before the crisis struck. What mainly characterizes Southern Europe in the decade leading up to the crisis is slow growth or stagnation combined with deteriorating competitiveness (manifest in unusually slow growth of exports).

Section 2.4 examines the development of the external balances of European countries in the decades leading up to the crisis in more detail. Distinguishing between countries in the South, the North, and the East fits the evidence quite well. The Southern countries have combined unusually slow growth (or stagnation) with increasing deficits on their current accounts. The Northern countries did not grow very fast, either, but their current accounts have improved year by year—the opposite of the Southern pattern. Hence, there appear to be important differences in the underlying dynamics. The Eastern countries also have a distinct pattern, combining relatively rapid growth in the years leading up to the crisis with large external deficits. These deficits, however, were primarily a consequence of high investment rates, not excessive wage (or consumption) growth, and should not necessarily be regarded as a matter of concern in a long-run perspective. Nevertheless, due to their increasing dependence on foreign capital, the Eastern countries were hit hard when the financial crisis struck.

In section 2.5 we delve deeper into the consequences of globalization and European economic integration during the decades leading up to the crisis for economic performance in the North, East, and South of Europe, drawing on data from the World Input-Output Data Base (WIOD).[2] The *raison d'être* of European integration efforts during the last decades, such as the deepening of the 'internal market' and the adoption of a common currency, has been that it would boost intra-European trade and growth all over Europe. We demonstrate, however, that the effects have been quite asymmetric. The major beneficiaries of increased European economic integration have been the previously socialist economies in the East, although the Northern countries have also benefited substantially. However, the Southern countries have benefited little, if at all, from this process.

Finally, section 2.6 draws the lessons from the analysis and considers the implications for policy. We argue that the present insistence on austerity, with economic stagnation, high unemployment (especially among the young), and increasing poverty as some of its main consequences, only adds to the problems caused by inconsistent and inefficient policies in the past (and present). What is needed is a European growth policy, properly adapted to the different capacities across Europe, which explicitly aims to diminish the large differences in productivity, income, and living conditions across the Union.

2.2. Global Transformations and Europe's Role, 1995–2011

Europe's growth is intimately interconnected with that of the global economy. In this section we consider how European countries have performed compared to other global players before and after the financial crisis struck. However, in order to assess the growth performance of European countries, we need a theoretical perspective that helps in the interpretation of the data. A useful approach, which will guide the analysis in this section, is the technology-gap theory of economic growth,[3] developed by economic historians such as Gerschenkron (1962), Abramovitz (1986), and Maddison (1991).

The technology-gap theory emphasizes that the capitalist world economy consists of countries with very different levels of economic and technological development, and that this provides each country with different opportunities and challenges. For example, countries far behind technologically and economically—such as China and India today—have the potential for rapid

[2] See Appendix for more information about the WIOD.
[3] For an overview, see Fagerberg (1994).

economic growth by exploiting technology already in use in more advanced parts of the world. Hence, provided that they manage to exploit this potential successfully, we would expect China and India to grow faster than, say, the US or Germany. Similarly, successfully exploiting this potential might allow the poorer countries in the East or South of Europe to grow faster than the richer ones in the Northern part of the continent.

However, historical evidence has shown that transforming potential growth into reality can be much more challenging than some of the early advocates of the approach tended to believe. Indeed, there have been big variations over time in the extent to which the potential for catching up—leading to convergence in productivity and income levels across countries—was exploited. Some periods witnessed considerable catching up, such as the decades following the end of the Second World War, which led Abramovitz to call this period the 'Catch up and convergence boom' (Abramovitz, 1994a, b). In other periods, however, catching up has been more or less absent. For example, very few developing countries managed to exploit the potential to reduce the gap with respect to the already developed part of the world during the 1980s and early 1990s, a period that has been characterized in the scholarly literature as the 'lost decades' for development (Easterly, 2001). Hence, the capacity to exploit the potential certainly cannot be taken as given and needs to be carefully explored. Abramovitz (1986) used the concept 'social capability' to characterize the different capacities of nations to exploit the opportunities for catching up economically and technologically. He also pointed out that the capabilities that are needed may differ depending on the technologies in question—successful exploitation of the potential for catching up depends on what he called 'technological congruence' (Abramovitz, 1994a)—and that the conditions for catching up therefore may vary over time.

Nevertheless, during the period from the latter half of the 1990s until the financial crisis struck, the global economy once again entered a more dynamic state, characterized by rapid catching up both globally and within Europe. To illustrate these developments, Figure 2.1 plots growth of GDP per capita between 1995 and 2008 against its level in 1995 for a sample[4] of countries at different levels of development. Although the main focus is on Europe and countries with which European countries may be most appropriately

[4] The country coverage is determined by the World Input-Output Database (WIOD), which is the main source of data in this chapter. WIOD provides time-series for forty countries worldwide, including all EU countries, and the rest-of-the-world from 1995 onwards. Although other data sources, such the World Bank, may have broader coverage, we have—largely for concerns about consistency—chosen to use the same sample throughout. Hence, except for section 2.4 (which is concerned with developments for subgroups of countries), all empirical analyses in this chapter are based on the same sample of countries (data for Taiwan lacking in a few cases). Note, however, that due to lack of space in the figures, not all observations included in Figures 2.1–2.6 and 2.12 have country names attached. See Appendix for more information on the data.

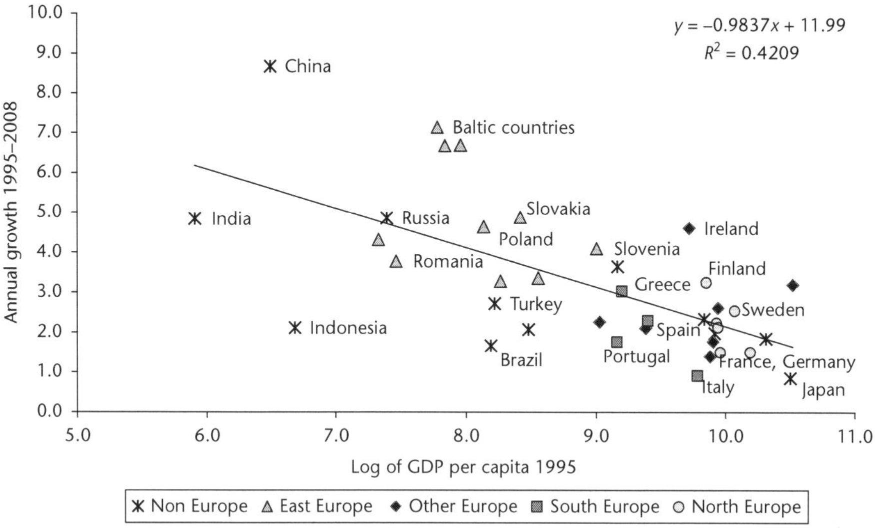

Figure 2.1. Growth and convergence in the global economy, 1995–2008
Source: Calculations by the authors using data from the World Bank (see Appendix for details)

compared, the sample also includes some major players from the developing part of the world (such as Brazil, China, India, and Indonesia). Thus, the countries included represent a major share of world GDP.

If poorer economies manage to exploit the potential for catch-up, as Gerschenkron's theory would suggest, growth rates should decline with rising GDP per capita levels. This is generally what is observed in the figure. If sustained over time, such a process would lead to reduced differences in levels of economic development between rich and poor countries—what in the economics literature has been termed 'convergence'. The convergence we observe in the period under study here stems from several sources. One of these is the much heralded rapid growth in the two largest developing nations in the world, China and India. This relatively new and very important trend has contributed to lifting hundreds of millions of people out of poverty. Another source of convergence is rapid growth in a number of former socialist economies in Europe, both in smaller countries such as the Baltic nations and larger economies such as Poland and Romania, to mention some important examples. Although, as expected from the theory, more developed countries tend to grow more slowly than average, there is also significant variation between them. For example, given their initial level of development, Ireland and Finland grew relatively quickly, while the opposite was the case for France and Germany. Some of the Southern European countries (indicated with squares in the figure) also grew slowly, particularly Italy.

The transformation of the global economy during this period was related to important technological, institutional, and political changes. The ICT revolution, which had been under way for a long time, started to have a major impact during these years. The period when computers, as Robert Solow once remarked, were visible everywhere except in the productivity statistics (Solow, 1987) was finally ended. In particular, ICT facilitated the coordination of economic activities across large distances (Hildrum et al., 2011). Together with major advances in transport technology, this presented firms—especially larger ones—with new opportunities for combining economies of scope and scale, serving markets worldwide and sourcing labour and other resources globally. Political changes in China and its gradual inclusion in the capitalist world system from the 1980s onwards added hundreds of millions of workers and (potentially even more) consumers to the global capitalist economy, as did similar changes in other parts of Asia in the following decades, with rapid growth as the result. The collapse of the Soviet empire and the fall of the Berlin Wall opened up not only new markets but also a vast pool of relatively skilled labour. The gradual inclusion of many former socialist countries into the European Union facilitated the transition and influenced the political, institutional, and economic changes in these countries. After painful transition periods (of varying lengths) rapid growth followed, making the period from the early 1990s to the financial crisis in 2007–8 perhaps most dynamic in Europe's recent economic history.

The financial crisis had a major impact on the global economy and Europe in particular. Figure 2.2 repeats the exercise from Figure 2.1, plotting growth of GDP per capita against its initial level, but now for the period after 2008.

The crisis clearly did not reverse every trend from the previous decades. China still grew very fast, and India and Indonesia grew even faster than before. Turkey and Brazil also kept growing. So the major global transformation pointed to here was still unfolding. Europe, in contrast, slid into recession, although the effects of the crisis were far from uniform. Among EU countries, only Poland did reasonably well during these years. Although six other European countries recorded positive growth rates of GDP per capita during this period, these were all in the 0–1 per cent range, very low in historical terms. There was a dramatic reversal of fortune for previous high-flyers such as Ireland, the Baltic countries, and Romania, and the Greek economy almost collapsed. Moreover, the slow-growing countries in Southern Europe—Italy, Spain, and Portugal—continued to underperform (with negative GDP per capita growth).

Thus, it seems that slow growth in Southern Europe compared to other countries at a similar level of development is not new. In the next sections we will delve deeper into the causes of this phenomenon. The decade before the crisis was characterized by rapid growth (i.e. catching up) in the Eastern

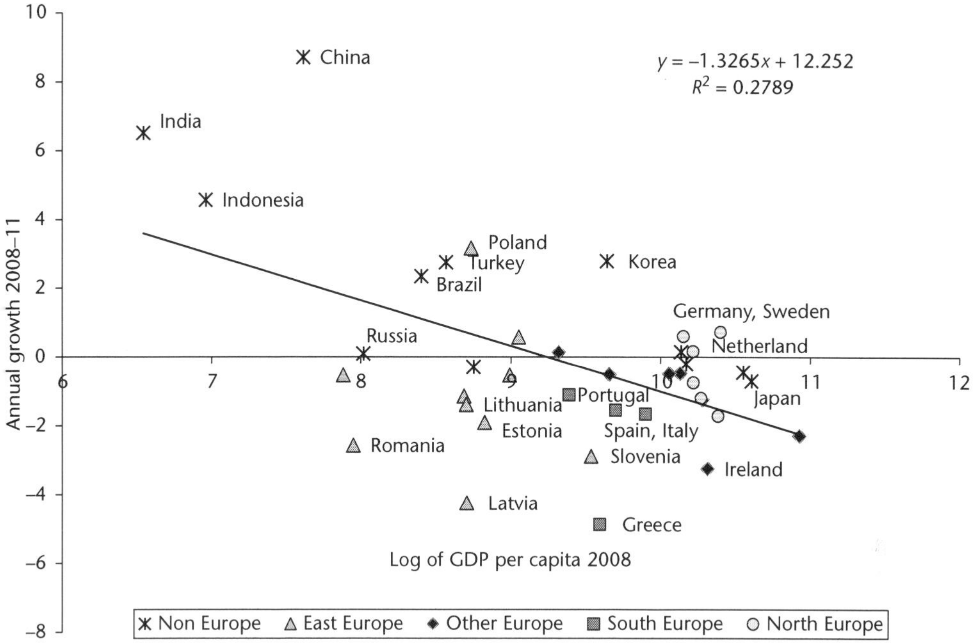

Figure 2.2. The crisis years, 2008–11

Source: Calculations by the authors using data from the World Bank (see Appendix for details)

part of Europe, but this process ran into trouble when the crisis hit. In general, the countries in the North of Europe have been less affected by the crisis than their counterparts in the South and the East. We will return to some of the reasons for these observations in sections 2.5 and 2.6.

2.3. Southern Exceptionalism?

As a first step towards understanding the causes for the stagnation in Southern Europe, we will consider the widely shared—but, as we shall show, erroneous—view that that this is the consequence of a deviant macroeconomic stance in these countries, for example allowing wages to grow much faster than productivity, thereby damaging cost competitiveness, or allowing consumption to grow much faster than GDP. The discussion draws on data from the World Input-Output Data Base (WIOD).[5]

[5] See Appendix for details.

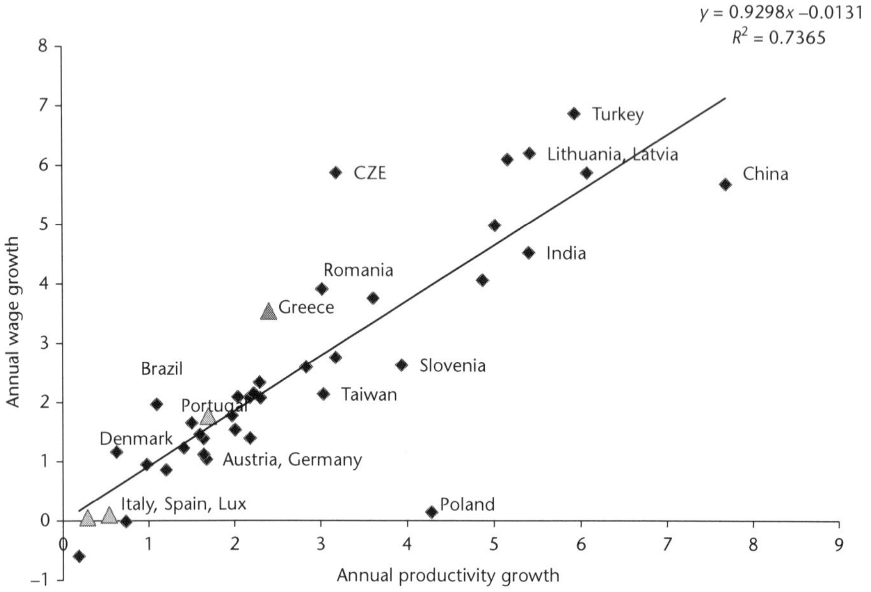

Figure 2.3. Growth of wages and productivity, 1995–2008

Note: Southern European Countries indicated by triangles

Source: Calculations by the authors based on data from WIOD (see Appendix for details)

Figure 2.3 plots growth of wages against growth of labour productivity—both in constant prices—from the mid-1990s to the outbreak of the financial crisis.

The figure confirms the prediction from growth theory that under capitalist conditions wages should grow in line with productivity (Pasinetti, 1974). Overall, the relationship is remarkably close,[6] although there are some clear outliers, with the Czech Republic above the regression line (with wages growing markedly faster than productivity) and Poland[7] and China below as the most obvious examples. Some other European countries also had relatively slow wage growth. Germany is the economically most important case, but it also holds for certain smaller countries (Austria, Slovenia, and Luxembourg). In fact, after the turn of the millennium, German wages grew very little if at all for a number of years, as a result of an effort by the government in agreement

[6] Note however, that the correlation coefficient is less than 1, consistent with a general tendency (albeit slow) towards increased inequality between wage earners and owners of capital, as observed by Piketty (2014).

[7] Poland is an extreme outlier. According to the WIOD data (which are based on national accounts), Polish wages hardly grew during the period under investigation here, while labour productivity grew around 4% per year on average. Other sources of information on Polish wages in this period indicate slightly higher growth but still substantially less than for labour productivity (Drozdowicz-Biec, 2013).

with business and trade union leaders to boost the country's competitiveness. For the three slow growers in Southern Europe, wage growth was either in line with (Portugal) or below (Italy and Spain) productivity growth. Only for Greece, which grew more rapidly than the other three in the period leading up the crisis (Figure 2.1), is there clear evidence of wage growth outstripping productivity growth.

Figure 2.4 carries out a similar exercise for the relationship between GDP growth and growth in private consumption. Again, the relationship between the growth of the two aggregates is remarkably close (as indicated by the high *R*-square) with China and Luxembourg as the most obvious outliers (relatively slow growth of private consumption when compared to GDP growth). However, in Greece, Italy, Portugal, and Spain (indicated by triangles) private consumption grew in line with GDP, as in most other countries.

However, there is more variation in public consumption, as Figure 2.5 indicates, with economically successful cases such as Turkey and Korea having much higher growth in public consumption than in GDP (or in private consumption for that matter), while the opposite is true for Germany, Slovenia, and the Baltic states. The Southern European countries (indicated by

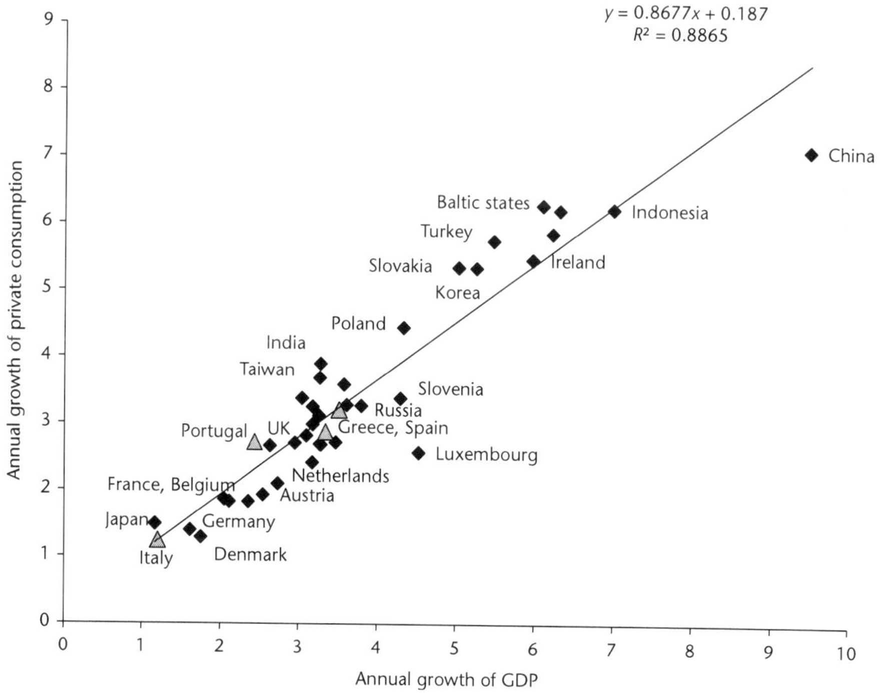

Figure 2.4. Growth of private consumption and GDP, 1995–2008

Source: Calculations by the authors based on data from WIOD (see Appendix for details)

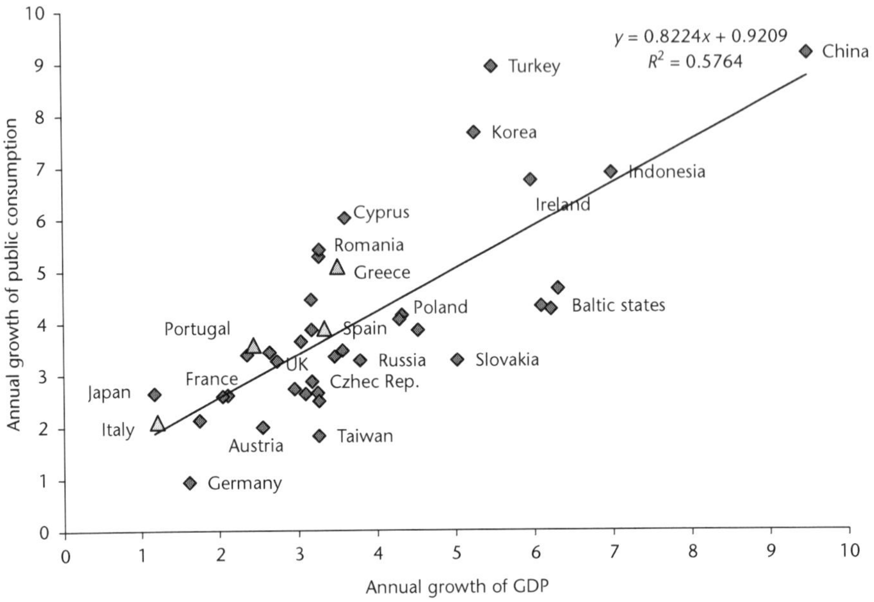

Figure 2.5. Growth of public consumption and GDP, 1995–2008
Source: Calculations by the authors based on data from WIOD (see Appendix for details)

triangles) tend to lie between the two extremes. However, in the Greek case, public consumption grew substantially faster than GDP.

Thus, with the possible exception of Greece, the growth patterns for wages and consumption in the Southern European countries do not deviate significantly from those of other countries. However, the three Southern slow growers and Greece do have another feature in common, which may be of much greater relevance in explaining their performance. This has to with their involvement in the global economy.

To examine this phenomenon, we look at the role of exports in GDP, using data from the World Bank. Figure 2.6 shows, on the horizontal axis, exports as a share of GDP. On the vertical axis, we plot the change of this share over 1995–2008. We see that, under the influence of globalization, the share of exports in GDP has tended to increase in almost all countries (the average increase is close to 10 percentage points).

As the figure shows, Greece, Portugal, Spain, and Italy (indicated by black squares) cluster together in the bottom-left quadrant,[8] this includes countries

[8] This, it may be noted, does not hold for Ireland which is often lumped together with some of the Southern economies as 'PIGS' countries. However, a detailed analysis of the Irish performance suggests that it has very little in common with the countries in the South of Europe and would warrant a separate analysis that we have chosen not to pursue here.

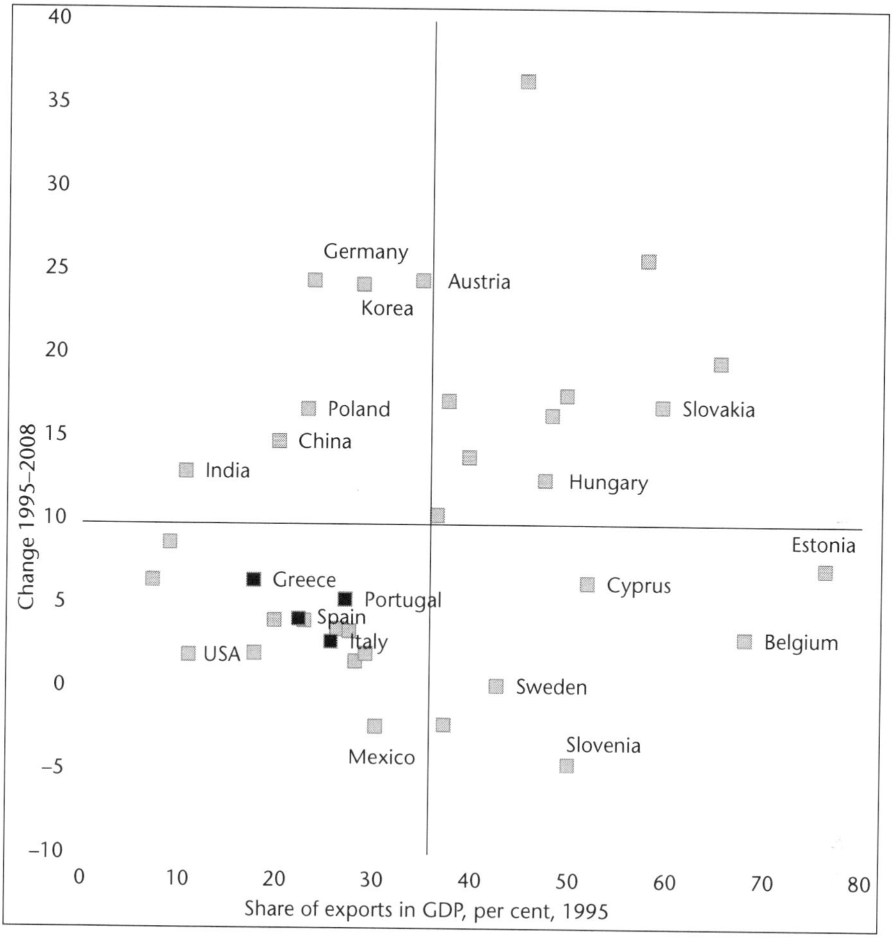

Figure 2.6. Exports as a share of GDP, level versus change, 1995–2008

Source: Calculations by the authors using data from the World Bank (see Appendix for details); Luxembourg (a major outlier) was excluded from the graph

that are below average both with respect to their initial involvement in trade[9] and the increase in the share over time. In contrast, countries such as Germany, Korea, and Austria, which did not differ much from the Southern European countries under study here with respect to the initial involvement in trade, recorded large increases in the share of exports in GDP. This also

[9] The degree of involvement in international trade is also a function of country size. Very large economies, with vast domestic markets, trade less externally than other countries. This explains why the USA, India, and China are to the left, while a very small country such as Malta is to the right in the figure. While this is to be expected, it is more surprising to find relatively small economies such as Greece and Portugal close to the US position. This tells us that, in relation to their size, the involvement of these countries in world trade is exceptionally low.

holds for several other countries that grew rapidly during this period. Hence, it appears that the Southern European countries have not managed to tap into the combined potential of technological change and increasing globalization to the same extent as many other countries. In the following, we will look more closely at the impact that this has had on the current accounts of these countries.

2.4. External Balance: The European Divide

In this section we analyse the development of the current accounts of selected European countries in the decades leading up to the financial crisis based on statistics from the IMF. The aim is to document that there are three clearly distinguishable categories in the form of the North, East, and South. For the sake of space, the analysis in this section will be limited to 21 European countries for which the division into these three groups is clearly relevant.[10]

A deficit on the current account means that a country's national debt increases. If sustained over several years, a country may enter a vicious circle of increasing debt and rising interest rates, and eventually face problems with respect to refinancing and servicing of its debt. The development of the current account is therefore of immediate interest here. Changes in the current account may be caused by changes in transfers (for instance, support from the EU), factor income (profit and interest, for example), and trade. However, what primarily drives the current account in the longer run is normally the trade balance.

It should be noted that rapidly growing countries often present foreign investors with favourable investment opportunities which they are eager to exploit. This makes it possible for a country to grow faster than it would otherwise have done, and will *ceteris paribus* lead to a deficit on its current account. Such deficits should not necessarily be regarded as a problem, however, because they are related to investments which will generate income in the future. Indeed, this is a normal aspect of catching up processes. When assessing the current accounts of European countries, it is necessary to take such factors into account.

Figure 2.7 displays the current account as a percentage of GDP prior to the crisis for six Northern European countries, of which four are euro-members and one pegs its currency to the euro (Denmark). As the figure shows, there was a general upward trend in the external balance for these countries. This is

[10] Numbers for the countries not included here will be reported in a more aggregated fashion in the next section as 'Other Europe'. Belgium, Cyprus, France, Ireland, Luxembourg, Malta, and the UK were included in this residual category. See also the discussion in section 2.6.

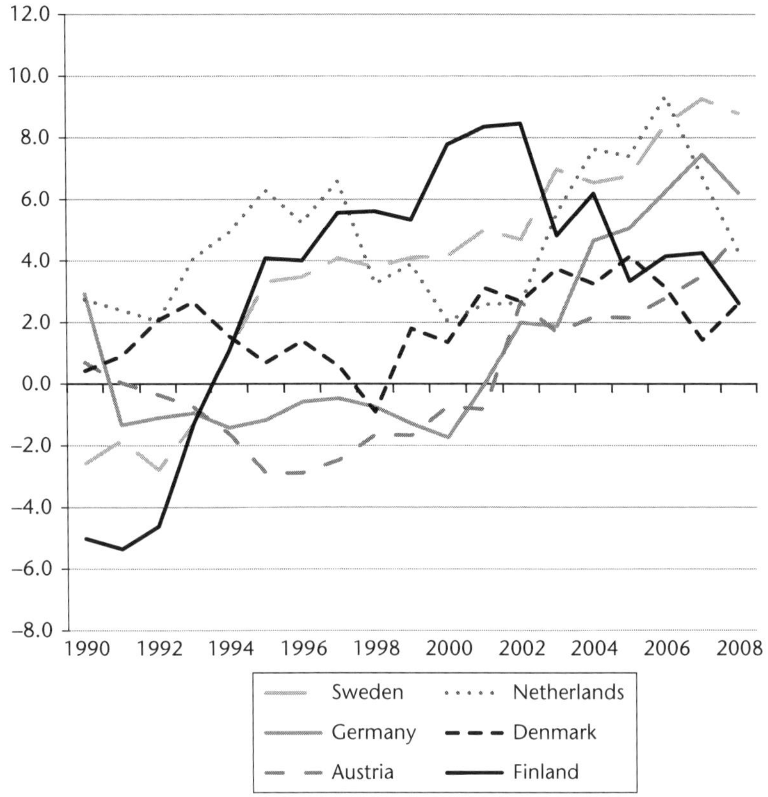

Figure 2.7. Northern Europe: current accounts, % of GDP, 1990–2008

Source: Calculations by the authors based on IMF data (see Appendix for details)

particularly evident from the period after the introduction of the euro around the turn of the millennium. Consequently, prior to the financial crisis all six countries had substantial surpluses on their current accounts.

The situation is quite different in Eastern Europe, as Figure 2.8 shows. Most countries in this group had deficits over the entire period, and these tended to increase, particularly after the introduction of the euro. As the figure shows, there is a clear difference between the 'Western' part of Eastern Europe, namely Slovenia, the Czech Republic, Slovakia, Poland, and Hungary, which quickly became embraced by the capitalist West, and the remaining countries, for which this process was slower.[11] Among the former, although most countries in this category continued to run deficits, these did not deteriorate over time. This is different from the latter group of countries, for which the deficits

[11] See Landesmann (Ch. 3 in this volume) for a more in depth analysis of the development of the economies in Eastern Europe.

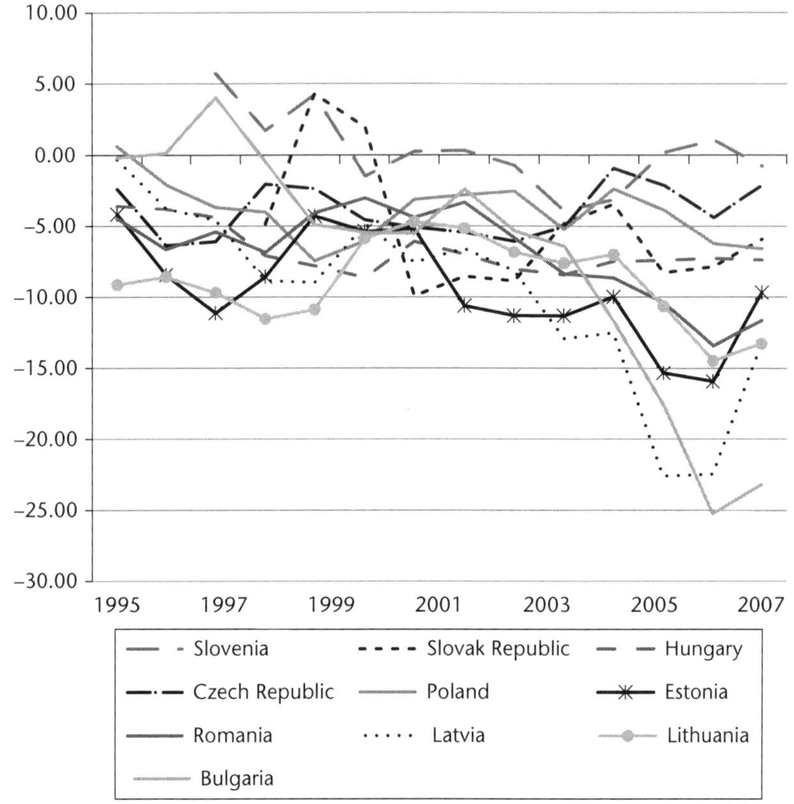

Figure 2.8. Eastern Europe: current accounts, % of GDP, 1995–2008
Source: Calculations by the authors based on IMF data (see Appendix)

continued to increase and eventually became very large (more than 20 per cent of GDP in some cases).

However, as already noted, these developments cannot be properly understood without taking into account the very high levels of investment in these countries during these years. As Figure 2.9 shows, in the first group of countries, investment was booming in the late 1990s, after which it levelled off. In contrast, for the second tier (the latecomers) investment only took off after the beginning of the millennium (see Figure 2.10). A veritable investment boom followed, with investment shares in GDP close to 40 per cent for some countries in certain years. Hence, the large deficits that emerged on the current accounts for countries such as Latvia, Estonia, and Bulgaria basically reflect an investment boom in which foreign investment played an important role. This does not mean that no domestic savings took place in these countries. In fact, according to calculations by the authors on WIOD data, the average domestic saving rate in these three countries between 2005 and

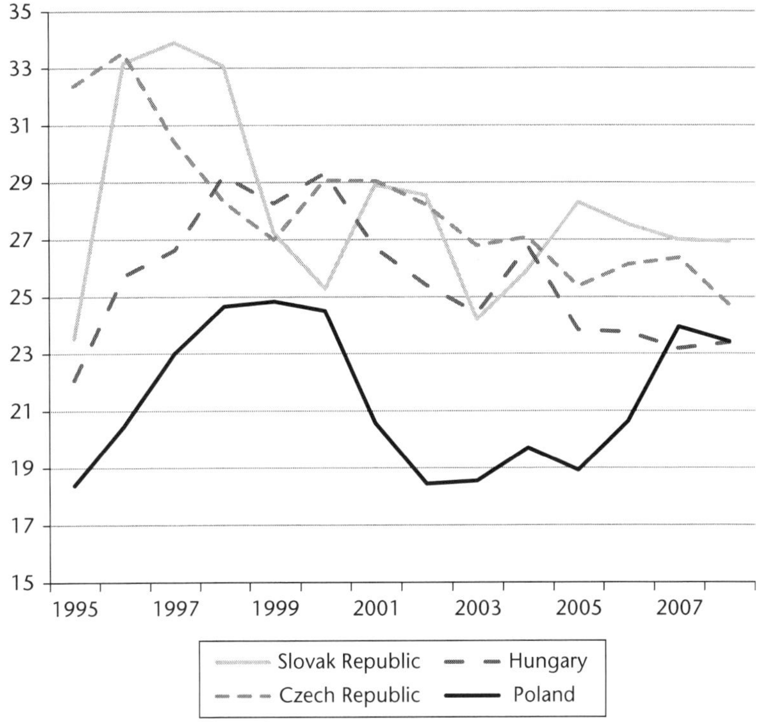

Figure 2.9. Early entrants: investment as % share of GDP, 1995–2008

Source: Calculations by the authors based on WIOD (see Appendix)

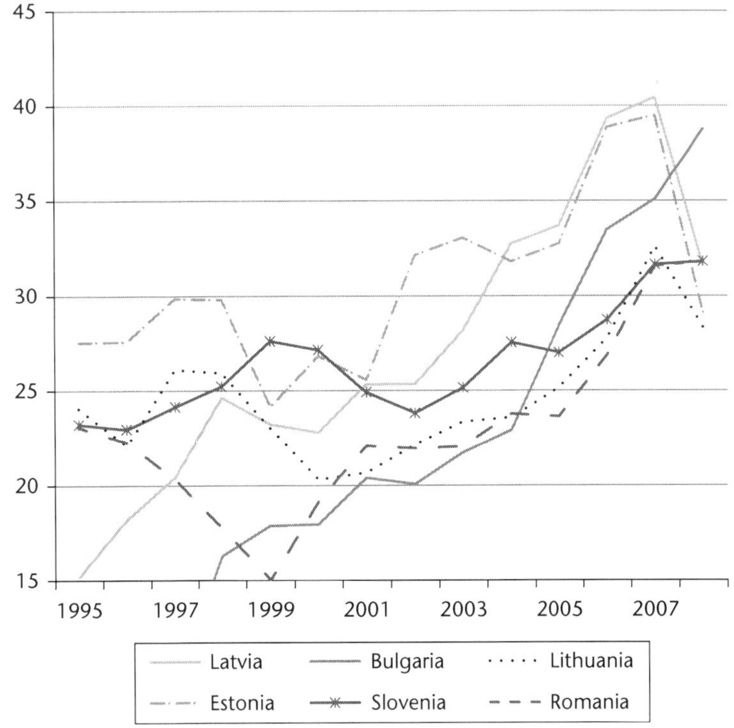

Figure 2.10. Late entrants: investment as % share of GDP, 1995–2008

Source: Calculations by the authors based on WIOD (see Appendix)

2008 was 18.5 per cent, and only for one country in one year did it slip below 15 per cent.

Figure 2.11 reveals similar evidence for Southern Europe (Italy, Greece, Portugal, and Spain). Around 1993–5 the current accounts of these countries were roughly in balance. Then a divergence appeared between Greece, Portugal, and Spain on the one hand, which moved into ever larger deficits from the mid/late 1990s onwards, and Italy on the other, which stayed in surplus for another five to ten years, after which deficits started to emerge there too. This apparently somewhat better performance of Italy on this dimension should not, however, be seen as an indication of superior competitiveness. It primarily reflects the fact that the rate of growth of GDP (see Figure 2.1), and hence also of imports, was slower in Italy than elsewhere.

A relevant question may be: 'To what extent does this pattern mirror the investment cycle, as in Eastern Europe?' The answer is: 'Much less so.' There was something resembling an investment boom in Portugal around the turn of the century and again in Spain closer to the end of the period, and these developments clearly influenced the situation in these two countries.

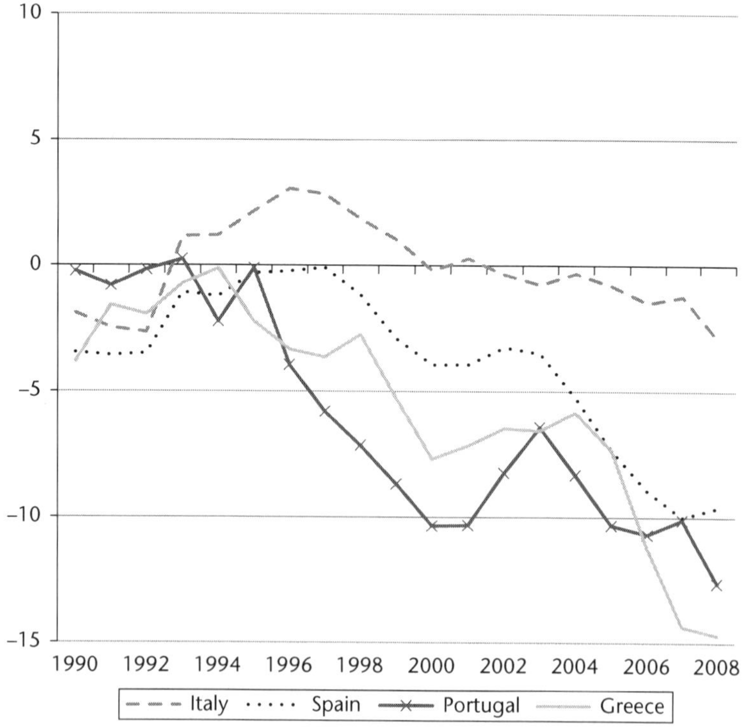

Figure 2.11. Southern Europe: sliding into current account deficit (% of GDP)
Source: Calculations by the authors based on IMF data (see Appendix)

However, this does not explain why Portugal's annual current account deficit continued to be close to 10 per cent (or more) even after the investment boom had faded, or the large and increasing deficits during the same period in Greece. Nor does the level or fluctuation in investments explain the more modest deficit that subsequently appeared in Italy.

Thus, slow growth and deteriorating current accounts over a number of years seem to be features that distinguish the countries in the South of Europe from those in the East or the North. What is the relationship between these trends and the process of European economic integration and the changes in the global economy from the early 1990s onwards? This is the question to which we now turn.

2.5. European Growth (or Lack of it): The Impact of European Economic Integration and Globalization

In this section we will delve more deeply into the roles of European economic integration and globalization for Europe's growth. We focus in particular on the performance of the three groups of countries identified—North, South, and East Europe. As before, the analysis is based on WIOD data. For comparison, the analysis again includes a group of 'other European countries', mostly from the Central and Western parts of Europe, and a group of non-European countries, including some large and economically important developing countries (China and India, for example).

We start by decomposing average annual GDP growth over the period 1995–2008 into two parts: one associated with trade (traded GDP,[12]) and the other associated with domestic sources. Figure 2.12 illustrates the contributions of the two categories of GDP to overall GDP growth for the forty countries included in the database.[13] The line stretching from the origin to the upper-right corner is the 45 degrees line. Below (above) this, the contribution of domestic (traded) GDP is largest.

The following patterns may be observed. The countries in Northern Europe cluster in the bottom-left of the figure. They grew relatively slowly, and with one exception the contribution from traded GDP exceeded that from

[12] See e.g. Johnson and Noguera (2012) for an elaboration on this concept.

[13] The contribution of traded GDP to overall GDP growth was calculated as the growth rate of traded GDP in a country multiplied by the initial (1995) share of traded GDP in total GDP (and similarly for domestic GDP).

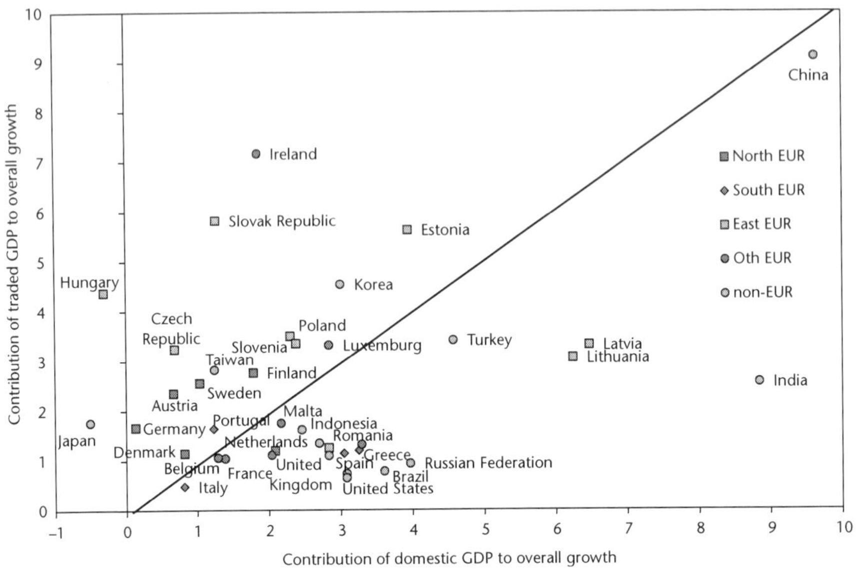

Figure 2.12. Contributions of domestic and foreign sources to growth of GDP, 1995–2008

Source: Calculations by the authors based on WIOD data (see Appendix); Bulgaria (a major outlier) was excluded from the graph

domestic sources. It may be noted that for Germany the contribution from domestic sources to growth was almost nil, reflecting the strict economic policies pursued in that country over a number of years.

The countries in the South of Europe are to be found in the bottom of the figure. Thus they too grew relatively slowly. Italian growth was particularly low. However, with the exception of Portugal, all Southern European countries have a larger contribution from domestic sources to growth than from traded growth, which clearly distinguishes them from their Northern neighbours.

In contrast to the stagnation elsewhere in Europe, many Eastern European countries grew quickly between 1995 and 2008, and in most cases they had a higher contribution from traded than from domestic GDP.[14] Their growth pattern thus does not resemble that of Southern Europe.

The group of 'other European countries' is more heterogeneous. Several countries (France, for example) are relatively similar to Southern Europe, in the sense of having slow growth and a high dependence on domestic sources. Ireland is different though, with high growth and a strong reliance on foreign sources.

[14] This holds for all countries in this group except Lithuania, Latvia, and Romania.

With respect to the non-European countries, it is interesting to observe that most of them have a lower contribution from traded GDP to growth than from domestic sources. The exceptions to this rule are Japan, Korea, and Taiwan.

The picture that emerges from this analysis of European growth is one of large differences across countries with respect their involvment in trade and the effects thereof. While some European countries are among the most internationalized in the world, in the sense of sourcing most of their growth abroad, other countries—notably those in Southern Europe—clearly do not belong to this category. These differences may seem surprising, given the efforts over the years to promote European growth and trade through the gradual elimination of barriers to trade (e.g. the internal market) and the introduction of a common currency, the euro. To gain a better understanding of the roles played by European economic integration and globalization for different parts of Europe, Figure 2.13 presents a decomposition of the GDP growth of the country-groups identified, which splits the foreign contribution according to the region where the final demand originated.

First, it may be noted that the effect on growth from trade with non-European countries (the area to the far right in the figure) exceeds that of intra-European trade for most country groups. The exception is Eastern Europe for which European demand has been very important indeed, supporting the rapid growth and 'catch-up' there from the 1990s onwards. Intra-European traded GDP accounted for 37 per cent of total GDP growth in these countries, which is about the same as the contribution from domestic sources. The countries in Northern Europe also received a substantial contribution from intra-European trade, 29 per cent of their total growth, which is larger than the contribution from domestic sources. In contrast, for Southern Europe the contribution to growth from intra-European trade is very small, only 11 per cent, and the largest part of this comes from intra-Southern European trade. In fact, the contribution to growth from trade with Northern Europe is essentially nil. Thus, although European integration has proven to be very fruitful for the rapid development of Eastern Europe from the early 1990s onwards, and also yielded substantial benefits for Northern Europe, the economic benefits for Southern Europe are much less obvious.

In the previous section, evidence on the development of the current accounts of European countries was discussed. It was shown that, while these were roughly in balance for most countries in the early 1990s, since then a very unbalanced development has taken place, with increasing surpluses in the North and corresponding deficits in the South and the East. As noted earlier, the trade balance is an important part of the current account and—at least in the longer run—determines the evolution of the current account and the country's international debt. It is of interest, therefore, to explore how European economic integration and globalization influenced the

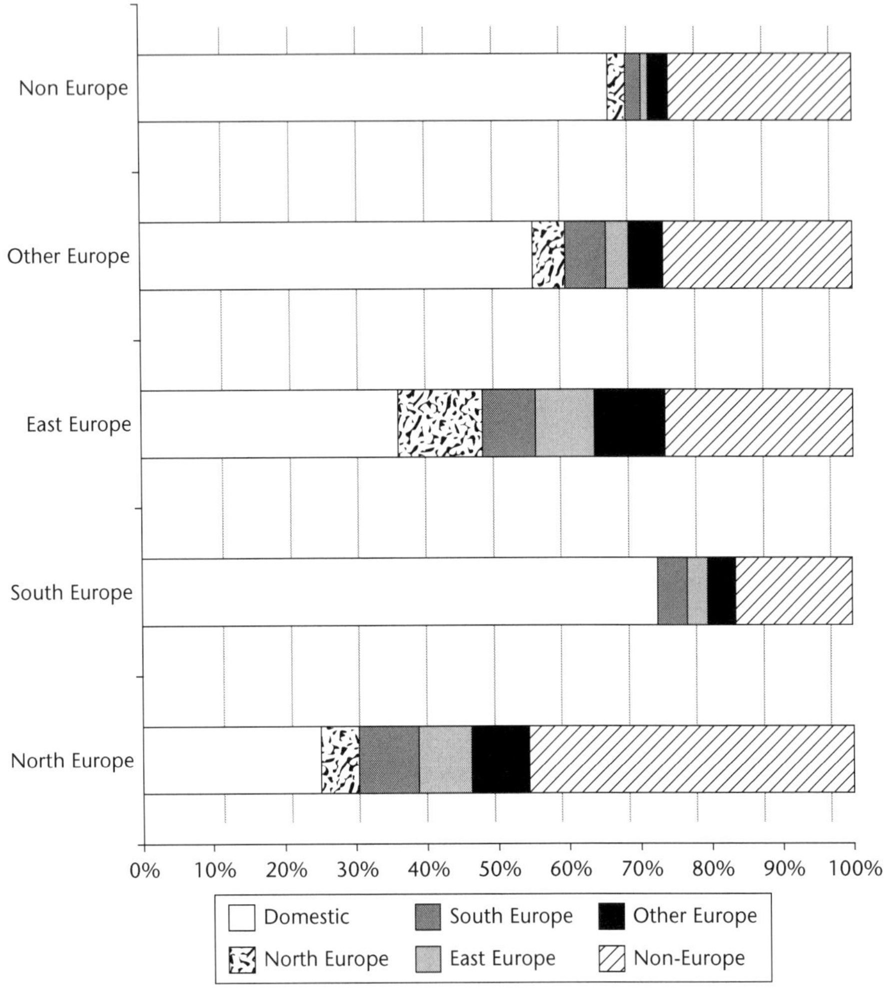

Figure 2.13. Decomposition of GDP growth, 1995–2008

Note: The regions to the left (vertical axis) are where the growth takes place, while the regions to the right (legends) are where final demand associated with this growth originates.
Source: Calculations by the authors based on WIOD data (see Appendix)

trade performance of different parts of Europe. To link the discussion of this issue more closely to the main topic of this section, namely the effects on growth, we look at what may be called the 'value-added trade balance' or the difference between 'exported GDP' and 'imported GDP'.[15] Figure 2.14

[15] This builds on the same methodology of decomposing value added (growth) by sources of demand as was used in Figures 2.12 and 2.13. Exported GDP of country *i* is produced in *i* but used (due to final demand) in a different country. Imported GDP in country *j* is GDP used (due to final demand) in *j* but produced in a different country.

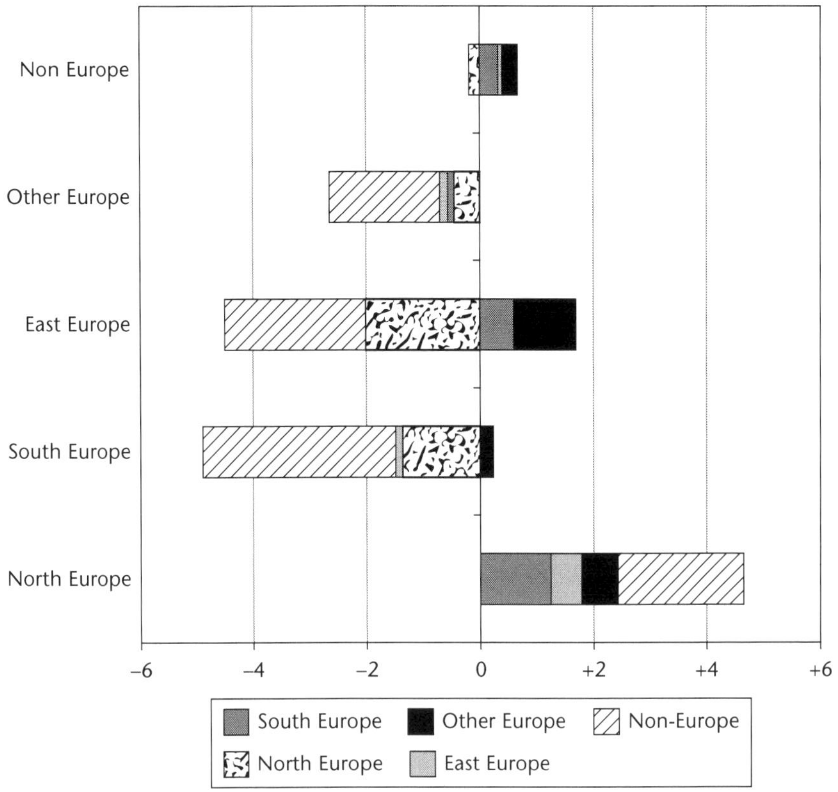

Figure 2.14. Change in the value-added trade balance (% of GDP), European regions, 1995–2008

Source: Calculations by the authors based on WIOD data (see Appendix)

illustrates changes in the value-added trade balance for various parts of Europe between 1995 and 2008, and how these changes were affected by trade with other parts of Europe and the rest of the world during this period.[16]

The imbalances that developed within Europe in the years leading up to outbreak of the financial crisis in 2007–8 are clearly visible in the figure. While, as pointed out, Northern Europe has improved its (value-added) trade balance by the equivalent of almost 5 per cent of GDP, the other European country-groups all saw their balances deteriorate, particularly in Eastern and Southern Europe. The factors leading to this deterioration were the same, namely increased deficits vis-à-vis Northern Europe, a clear beneficiary of European economic integration, and the rest of the world. However, in the case of Eastern Europe these negative effects were to some degree counteracted

[16] We use current prices to calculate this.

by gains vis-à-vis Southern and Other Europe. The decline in the balance was particularly large for Southern Europe, for which increasing deficits in trade with the rest of the world were the largest contributing factor. An important part of this loss was no doubt due to increasing competition from China and other industrializing countries. The Southern countries, especially Italy and Portugal, were vulnerable to increasing Chinese competition because their manufacturing industries were specialized in relatively mature, low-skill products, such as leather products, footwear, textiles, and clothing, all of which appeared as natural targets for producers in industrializing countries.[17]

This section has shown that, during the years leading up to the crisis, deepening European economic integration and globalization had very asymmetric effects on different parts of Europe. While Eastern Europe clearly benefited (at least until the financial crisis struck), and Northern Europe as well, the benefits are much less obvious for the countries in Southern Europe. In the next and final section we will discuss the implications of these findings for the analysis of the current stagnation in Europe.

2.6. Conclusions

Much of the European economy is currently characterized by stagnation, high unemployment, and—as a result—social problems, and this has been the situation for several years now. European policy-makers are busy creating new regulations, especially in the financial and macroeconomic policy areas, with the hope that Europe's problems will eventually go away. We think it highly unlikely that such regulatory fixes, however sensible they may seem, will be enough to restore Europe's growth and lead to reduced differences between richer and poorer parts of the union. The root of the problem is that Europe is a highly heterogeneous economic entity. Although the challenges that different parts of Europe face may to some extent be related, their capacity to deal with such challenges, and therefore also the economic and social outcomes, varies widely. Without taking such differences into account, any discussion of solutions to the present crisis runs the risk of ending up with unrealistic and ineffective proposals that may make the situation worse.

The evidence considered in this chapter suggests that there are three broad categories of countries on the European continent: North, East, and South.

[17] The increases in Chinese exports between 1995 and 2008 were concentrated in relatively mature, low-skill industries. As an example, consider the sector 'Leather, leather products, and footwear' in which China had a gain of 25.3% of the global market (world traded GDP), the strongest Chinese gain in any sector during this period. Italy and Portugal were among the main losers in the sector.

The Northern countries are advanced economies with high levels of productivity but which have tended to grow relatively slowly. They have taken advantage of the opportunities offered by European economic integration and globalization, and have internationalized their economies at a relatively rapid rate. As a result they have been running surpluses in their trade with other European countries and with the rest of the world. The financial crisis represented a serious challenge for these countries but no more than in other advanced economies.

The Eastern countries for historical reasons have productivity and income levels well below the world frontier, but relatively well-educated labour forces. They went through rapid economic, institutional, and political change, attracted considerable foreign investment, and increased their trade with other European countries and the rest of the world. Some countries joined this process early, while others only entered later. This resulted in rapid growth in this period but also external deficits and heavy dependence on international financial markets. These countries, and particularly the late entrants, were therefore hard hit by the financial crisis and the setbacks that followed, and their economies consequently contracted much more than Northern countries.

The Southern countries have productivity and income levels comparable to the most advanced Eastern economies. However, on average they grew much more slowly in the period up to the crisis. They are among the least internationalized economies in Europe, and have not managed to increase their presence in foreign markets to the same extent as most other European countries (or emerging economies from other parts of the world). A characteristic feature of several Southern countries is increasing external deficits and rising foreign debts over a number of years. As a result, they were particularly hard hit by the financial crisis and their economies contracted more than other European countries, with fast rising unemployment as one of the consequences.

These categories are not set in stone, however, and some reservations may well be in order. First, there is obviously a lot of variation within these categories, as countries differ in a number of respects, although we will not elaborate further on this issue here. Second, some countries have characteristics that would place them somewhere between the 'Northern' and 'Southern' archetypes. This holds, for example, for some of the countries in the Centre-West part of Europe such as the UK, France, and Belgium. Thus, some of the challenges that the Southern countries face may to varying degrees be relevant for other European countries as well.

The analysis presented in this chapter has shown that, because of Europe's heterogeneous character, the effects of globalization and European economic integration have been very different for the three groups of countries considered here. For Northern Europe, the classic advantage of integration, namely a larger market for the products that these countries are specialized in (in many cases, advanced, high-value-added products), seems to have contributed to an improved external trade balance and faster economic growth. For Eastern Europe, integration into the European economy allowed these economies to tap into some of the advantages of global economic integration, benefiting, for example, from relocation of production from the Northern European countries. For Southern Europe, however, globalization and European economic integration appear to have been much less beneficial, mainly due to a lack of competitiveness, which has also rendered them vulnerable to increasing competition, both at home and abroad, from China and other developing countries. As shown in the preceding section, about three-quarters of the increase in Southern Europe's trade deficit before the financial crisis struck came from trade with non-European countries, the remaining from increased deficits vis-à-vis the Northern countries.

The introduction of the euro has also had a markedly asymmetric impact. For Germany it dampened the upward drift of its currency that otherwise might have resulted from the increasing external surpluses that gradually built up as a consequence of the strict economic policies pursued by successive German governments. This occurred because Germany now shared a currency with a number of less competitive economies, such as the Southern countries and France. The German policy stance, however sensible it might have looked from a national point of view, gradually became a problem for the less competitive countries of the Eurozone, because it effectively meant that there was hardly any growth in the demand for their exports in Northern markets. This naturally added to the increasing competitive challenge these countries were facing from rapidly industrializing economies outside Europe, in particular China. In effect, for the Southern European countries, the euro merely removed a convenient mechanism to compensate for their lack of competitiveness without simultaneously hurting domestic demand and employment. For a while it allowed them to finance the growing external deficits at favourable interest rates, but with hindsight this proved a decidedly mixed blessing.

In the case of Southern Europe, being less competitive and encumbered by slow productivity growth, policy-makers were left with two main choices. Either they could restrain growth of domestic demand, leading to low growth in GDP, increasing unemployment problems (especially for newcomers to the labour market), and exerting a downward pressure on wages. This policy, while leading to stagnation, had the advantage that, mainly by curbing imports, it put a brake on the tendency towards increasing trade deficits with other countries. Or they could allow domestic demand to grow sufficiently to compensate for low demand from abroad and lack of competitiveness, preventing unemployment problems from getting out of hand, but at the expense of increasing deficits in their trade with other countries. Among the Southern economies, Italy was for a long time close to the first scenario, while developments in the remaining countries were closer to the second. What happened after the financial crisis struck, however, was that the second scenario no longer represented a realistic option. Thus, after a while, policy-makers in all Eurozone countries increasingly faced pressures to adjust policy towards a more austere stance, leading to stagnation and high unemployment in the entire Eurozone and in the rest of Europe as well. The consequences have been most strongly felt among the economically and technologically least developed, and hence the least internationally competitive, countries in the South and East of Europe, but the effects are visible in some countries in the North as well.

It has often been suggested that the governments and populations in Southern Europe behaved irresponsibly in the years leading up to the financial crisis. Whatever the merits of this view, the creation of a common currency for an economically extremely diverse continent, without any mechanism to coordinate economic policies and prevent problems from getting out of hand, appears to have been a much more irresponsible policy. Sadly, the present insistence on austerity, with economic stagnation, high unemployment (especially among the young), increased poverty, and social unrest as some of its main consequences, only adds to the problems created by a flawed currency arrangement, inconsistent macroeconomic policies, the lack of (constructive) policy coordination and leadership at the European level, and the failure to take the heterogeneous character of Europe's economy sufficiently into account.

Arguably, what is needed at the European level is a growth policy, properly adapted to the different contexts and capacities across Europe, which explicitly aims to diminish the large differences in productivity, income, and living conditions across the Union. But, as pointed out in the Introduction to this volume, to be sustainable the growth that such a policy leads to has to be of a different kind than in the past. Rather than increasing welfare and employment by burning fossil fuels, it has to fulfil these goals while, at the same time, substituting fossil fuels with renewable energy sources, i.e. nothing less than a fundamental transformation of technology, the energy sector (see Chapters 4 by Geels and 7 by Lauber and Jacobsson), and ways of life, as emphasized by Laestadius in Chapter 6 of this volume. Nevertheless, such a transformation, if carried out with the force necessary to achieve its aims, will require large public and private investments, leading to higher economic activity and the creation of new jobs, and helping to lift Europe out of its current stagnation. Hence, combining growth and sustainability is not impossible, but on the contrary the most promising way for Europe out of the crisis, as argued by

Mazzucato and Perez in Chapter 9 of this volume, although, as they point out, a substantial amount of policy innovation may be required.

Acknowledgements

Earlier versions of this chapter have been presented at the 'The Challenge for Europe in a New Age' workshop, Ålborg, Denmark, 14–15 March 2013, the Joint UNU-MERIT/ School of Governance Seminar, Maastricht, the Netherlands, 20 May 2014, and the 15th Conference of the International Schumpeter Society, Jena, Germany, 27–30 July 2014. We thank participants at these events, the editors of this volume and Witold Kwasnicki for useful comments and suggestions, retaining sole responsibility for any remaining errors or omissions.

References

Abramovitz, M. (1986). 'Catching up, Forging Ahead, and Falling Behind', *Journal of Economic History*, 46: 386–406.

Abramovitz, M. (1994a). 'The Origins of the Post-War Catch-up and Convergence Boom', in J. Fagerberg, B. Verspagen, and N. von Tunzelmann (eds), *The Dynamics of Technology, Trade and Growth*, Aldershot: Edward Elgar, 21–52.

Abramovitz, M. (1994b). 'Catch-up and Convergence in the Post-War Growth Boom and After', in W. J. Baumol, R. R. Nelson, and E. N. Wolf (eds), *Convergence of Productivity: Cross-National Studies and Historical Evidence*, Oxford: Oxford University Press, 86–125.

Drozdowicz-Biec, M. (2013). 'Why are Poles Poor Despite Working Hard?, <www.obserwatorfinansowy.pl>, accessed Nov. 2013.

Easterly, W. (2001). 'The Lost Decades: Explaining Developing Countries' Stagnation in Spite of Policy Reform 1980–1998', *Journal of Economic Growth*, 6: 135–57.

Fagerberg, J. (1994). 'Technology and International Differences in Growth Rates', *Journal of Economic Literature*, 32: 1147–75.

Gerschenkron, A. (1962). *Economic Backwardness in Historical Perspective*, Cambridge, Mass.: Belknap Press.

Hildrum, J. M., Ernst, D., and Fagerberg, J. (2011). 'The Complex Interaction between Global Production Networks, Digital Information Systems and International Knowledge Transfers', in C. Antonelli (ed.), *Handbook on the Economic Complexity of Technological Change*, Cheltenham: Edward Elgar Publishing, 395–416.

Johnson, R. C., and Noguera, G. (2012). 'Accounting for Intermediates: Production Sharing and Trade in Value Added', *Journal of International Economics*, 86: 224–36.

Maddison, A. (1991). *Dynamic Forces in Capitalist Development: A Long-Run Comparative View*, New York: Oxford University Press.

Pasinetti, L. (1974). *Growth and Income Distribution*, Cambridge: Cambridge University Press.

Piketty, T. (2014). *Capital in the Twenty-First Century*, Cambridge, Mass.: Belknap Press.
Solow, R. (1987). 'We'd Better Watch out', *New York Times Book Review*, 12 July.
Timmer, M.P. (2012). *The World Input-Output Database (WIOD): Contents, Sources and Methods*, WIOD Working Paper, 10 <http://www.wiod.org/publications/papers/wiod10.pdf>.

Appendix: Data and Sources

In Figures 2.1 and 2.2: GDP per capita is gross domestic product divided by midyear population. GDP is the sum of gross value added by all resident producers in the economy (plus any product taxes and minus any subsidies not included in the value of the products). It is calculated without making deductions for depreciation of fabricated assets or for depletion and degradation of natural resources.

In Figure 2.6: exports of goods and services represent the value of all goods and other market services provided to the rest of the world. They include the value of merchandise, freight, insurance, transport, travel, royalties, license fees, and other services, such as communication, construction, financial, information, business, personal, and government services. They exclude compensation of employees and investment income (formerly called factor services) and transfer payments. Source: World Development Indicators, World Bank, 28 September 2012, <http://data.worldbank.org/products/wdi>.

In Figures 2.7, 2.8, and 2.11: the current account is all transactions (between an economy and the rest of the world) other than those in financial and capital items. The major classifications are goods and services, income, and current transfers. Source: International Monetary Fund, World Economic Outlook Database, April 2012, <www.imf.org/external/pubs/ft/weo/2012/01/weodata/index.aspx>

For the remaining figures the World Input-Output Database (WIOD) provides a time-series of world input-output tables for forty countries worldwide and a model for the rest of the world. The database has been constructed on the basis of officially published input-output tables in conjunction with national accounts and international trade statistics. Countries included in WIOD are: Australia, Austria, Belgium, Brazil, Bulgaria, Canada, China, Cyprus, Czech Republic, Denmark, Estonia, Finland, France, Germany, Greece, Hungary, India, Indonesia, Ireland, Italy, Japan, Korea, Latvia, Lithuania, Luxembourg, Malta, Mexico, Netherlands, Poland, Portugal, Romania, Russian Federation, Slovak Republic, Slovenia, Spain, Sweden, Taiwan, Turkey, United Kingdom, and United States. When the analysis was conducted (January–March 2013), WIOD included data between 1995 and 2009. More recently it has been updated to 2011. For more information see: Timmer (2012). Source: <www.wiod.org/new_site/data.htm>.

3

The New North–South Divide in Europe: Can the European Convergence Model be Resuscitated?

Michael A. Landesmann

3.1. Introduction

Although global in character, the financial and economic crisis of 2008–14 is destined to be seen as a pivotal event in the history of Europe's development, particularly with respect to its cross-country integration experience and policies. In this chapter the focus is on the implications that the crisis has for the low- and medium-income economies of Europe, comprising the countries of Central, Eastern, and Southeast Europe (CESEE) and the Southern EU economies (Greece, Portugal, and Spain) as well as other countries in the EU's neighbourhood. For simplicity, I shall refer to this group of economies as 'Europe's South' in order to emphasize development gaps and imbalances in European integration.

By mid-2014 (when this chapter was being completed) it looks as if the worst of the euro-crisis is over, the worry of a dramatic meltdown of the European banking system has receded and a modest recovery is under way. On the other hand, European banks are still very weak and the move towards a 'European banking union' points to a glass less than half-full. Similarly the overall debt situation in many European economies is problematic regarding both the debt levels of the private-sector (household and corporate) as well as sovereign debt; the latter has considerably worsened compared to before the crisis because of the burden the banking crisis has imposed on many states, together with the impact of economic recession on fiscal balances and very weak economic growth. The European economy is thus not in a very strong state, while forecasts suggest a significant hiatus in the medium-term path of potential growth.[1]

The assessments of developments in 'Europe's South' have also become gloomier for the medium and longer term: we have seen dramatic declines in GDP during the crisis in the EU South. In Eastern Europe the economic crisis has also taken a severe toll: with the exception of Poland and Albania, most Eastern European countries experienced negative GDP growth rates over the period 2008–13; only a few countries had by the end of 2013 reached GDP levels exceeding those prior to the crisis (see Figure 3.2). Furthermore, the gains made in bringing down unemployment over the period 2000–8 in the CESEE region have, for the most part, been lost and in the EU South unemployment stands at record levels.

The underlying driver of the 'New North–South Divide' in Europe is, as we shall see, the build-up of external imbalances prior to the crisis both within the EU and in countries in Southeast Europe closely connected with the EU. The causes of this build-up will be discussed, as will the inadequacy of the institutional and policy framework of the EU and the Eurozone in particular. (Examples of contributions regarding external imbalances in the Eurozone are Darvas, 2012, and De Grauwe, 2012.)

The analysis in the chapter arrives at the following conclusions in the context of slowdown in growth and the emerging policy framework. The main differentiating feature among the emerging European economies that the analysis singles out is the pre-crisis build-up of (structural) current account disequilibria, associated developments in external debt, and the debt positions particularly in the private-sector (households and corporations). The pre-crisis build-up of disequilibria and debt accounts for most of the differentiated impact of the crisis on different economies over the period 2008–13.

A subgroup of three Central European economies (the Czech Republic, Poland, and Slovakia) has been scarcely affected by the debt build-up. These countries showed little sign of competitiveness problems in their tradable sectors (which is also the case with Hungary), while the EU South and most of the countries in Southeast Europe and the Baltic states developed unsustainable disequilibria in both these respects.

[1] Thus international forecasts (e.g. European Commission, 2013; IMF, 2014) see the European economy continuing to operate below potential for the next couple of years, implications for persistently high unemployment rates and hysteretic effects on longer term growth (discouragement effects on activity rates, skills erosion, reduced levels of infrastructure spending, less venture finance, and R&D expenditure, etc.). DG EcFin (see European Commission, 2013, table I.2) estimates in its central ('no policy change') scenario the potential output path of the euro-area for the period 2014–23 to be a mere 1.1% p.a. and that of GDP per capita 0.9% (the estimates for the US over that period are 2.5% and 1.8% respectively). This would imply a considerable income gap opening up between the US and Europe over that period.

Important groups of economies, such as the EU South, most of the countries of Southeast Europe, and some countries in Central Europe (Slovenia, Hungary), have entered a vicious circle: high initial debt levels and dim growth prospects translate into greater doubts about sustainability and hence higher interest rates that impose a constraint on investment and encourage corporate and household deleveraging (further compounded by the weak state of the banking system). This dampens consumption expenditures, and leads to cutbacks in employment (and wages) and, in turn, to lower household incomes and domestic sales prospects. The consequent lower growth prospects, again, raise concerns over debt sustainability and the need to keep interest rates high.

Prospects of offsetting factors such as a potential rise in competitiveness and hence export-led recovery are dim in the current context of low growth in the European economy as a whole.

The chapter thus points towards a sustained period in which the income convergence that characterized the decade prior to the current financial and economic crisis will either not proceed or will proceed at a much reduced pace. Deleveraging processes, difficulties in dealing with the large debt positions of the private sector, the weak banking system and persistence of national fragmentation of financial markets in Europe, as well as the associated feedback effects on sovereign debt, characterize many of the lower-income economies in Europe. The driving force of foreign direct investment and the build-up of cross-border production networks, which played a crucial role in enabling some of the successful Central European economies to build up and upgrade export capacities, also show weaker momentum than before the crisis (see Hunya and Schwarzhappel, 2013). Adjustment processes to deal with the pre-crisis neglect of a viable tradable sector and the need to build up adequate and modernized export capacities will have to be given greater priority. Consequently, the use of a range of policy instruments (particularly in the areas of training, labour market, and industrial and regional policies) focused on this task will need to be strengthened.

3.2. The European Growth and Convergence Model Prior to the Crisis

The pre-crisis integration model in relation to Europe's Emerging Economy (EME) region[2] was characterized by a high degree of liberalization of external

[2] We shall use the term 'EME' interchangeably with 'Europe's South'.

economic relations. Trade relations were thoroughly liberalized and there was a commitment to free international capital movement (in all its forms). In the CESEE region in particular, financial markets were fully opened up to foreign financial institutions and in most of these economies foreign banks attained a dominant market position (see e.g. Becker et al., 2010).

As Figure 3.1 shows, the period from the mid-1990s onwards coincided with a process of 'convergence' in many countries of the CESEE region, as these economies embarked (after a difficult first phase of 'transition') on a growth path with rates substantially above those of their Western neighbours. For a number of these economies, the 'catching-up process' was nonetheless interrupted at times by policy mistakes (such as misguided forms of privatization, and often problematic steps taken in monetary and exchange rate policy, in particular opting too early for a fixed exchange rate regime). The performance of the EU Southern economies reveals much less evidence of 'convergence' over this period, with Italy showing particularly low growth rates,[3] and both Greece and Portugal roughly maintaining their gaps in income levels relative to the EU as a whole, while only Spain experienced above average growth.

Underlying the growth performance of the CESEE economies was the opportunity which any lower-income, lower productivity economy has to benefit from 'technology' transfer (the so-called Gerschenkron effect.[4]) In the particular case of former transition economies, 'technology' should be interpreted rather broadly, including the importance of product design, organizational structures, and behavioural practices, facilitated by changes in institutions and in legal frameworks. In the case of many of these economies the speed of 'technology' transfer was reinforced by the anchoring to EU pre-accession and then accession arrangements. This anchoring added to the attraction of the region to foreign direct investments, a major conduit for the type of technology transfer already alluded to. Low relative unit labour costs combined with relatively high human capital endowment (adjusting for relative income levels) made the region attractive to foreign investors. This in turn led to access to high-income markets and the possibility of integration into cross-border production networks.

As will be shown in the next section, only in some of the economies did this lead to a substantial recovery of industrial production capacities, i.e. a process of 'reindustrialization' after the earlier period of—often massive—deindustrialization which most countries experienced at the beginning of the transition period. In many other economies, a longer period of political

[3] Italy is difficult to categorize as being a member of the EU's 'South' or 'North', as it has features of both.

[4] Named after Alexander Gerschenkron's 'advantage of backwardness' thesis; see Gerschenkron (1962).

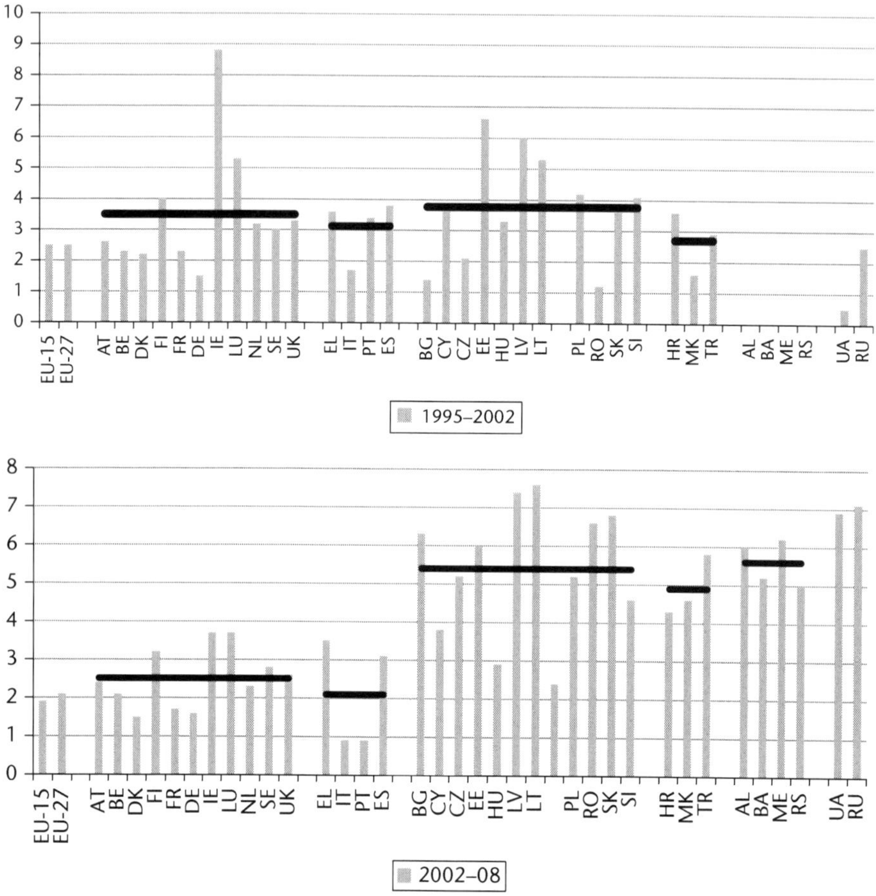

Figure 3.1. Growth of GDP, average annual growth rates, % (1995–2002, 2002–8)
Source: wiiw Annual Database incorporating national statistics, Eurostat. The lines represent simple arithmetic group averages. For explanation of country codes see page xix.

and economic turbulence, such as in most countries of Southeast Europe and in the Baltics, led to a situation in which pre-transition levels of industrial production were never attained; this, in turn, showed up in sustained gaps in trade balances. This had grave consequences in terms of vulnerability to external shocks, a point to which I shall return.

Furthermore, the economically weaker and vulnerable countries (in the Balkans and the Baltics) in particular adopted various versions of fixed exchange rate regimes. The reason was often a lack of trust in domestic monetary authorities and to avoid the large exchange rate fluctuations that can characterize shallow foreign exchange markets. By pegging the exchange rate, the countries also wanted to speed up financial and monetary integration

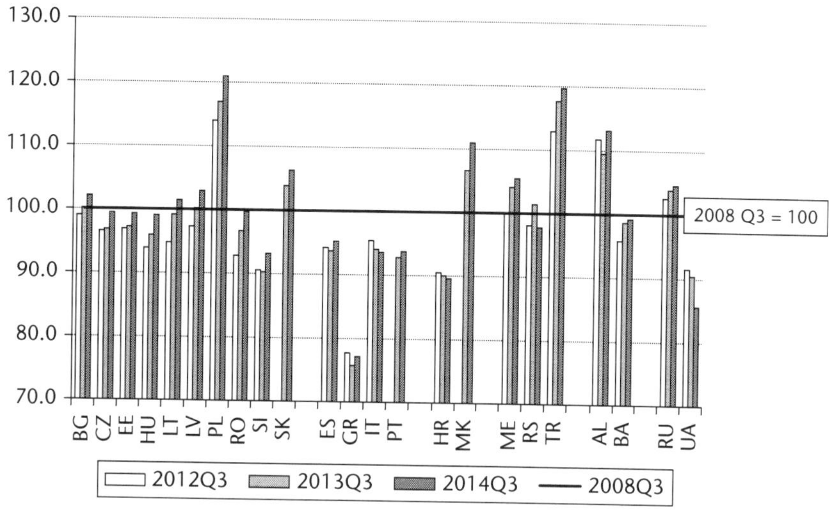

Figure 3.2. GNP levels third quarter 2012, 2013, 2014 (levels relative to 2008 = 100)
Source: Eurostat, national Statistics. For explanation of country codes see page xix.

with the euro-area. In turn, the choice of exchange rate regime contributed strongly to sustaining and accentuating the problem of deteriorating trade balances.

3.3. External Imbalances of Different Groups of Europe's Low- and Medium-Income Countries

The pre-crisis European integration framework (with its monetary and financial markets dimension) was designed to encourage large inflows of foreign investment from the more developed to the less developed countries, with external imbalances expected temporarily to widen, then to narrow and eventually close as income levels converged, mainly on account of export growth. The outcome in Central Europe (particularly the Czech Republic, Hungary, Slovakia, and Poland) has been more or less as intended, but not in the countries of Southern Europe nor in the Baltic states or in Southeastern Europe. The development of external imbalances prior to the crisis led to an accumulation of foreign debt with clearly unsustainable growth dynamics. Uncertainties surrounding the future growth of exports of goods and services point to a structural problem that may also restrain the growth of those countries in the longer term, given the somewhat unfavourable climate for foreign investment that has been developing (see Hunya and Schwarzhappel, 2013).

Groups amongst Europe's Low- and Medium-Income Countries

In the following discussion of economic developments in low- and medium-income countries in the European economy, we shall group countries on the basis of common and related features, the most important of which date back to pre-crisis developments. Particular emphasis is placed on the build-up of external and internal disequilibria prior to the financial crisis, which then led to different adjustment pressures following the onset of that crisis.

Figure 3.3a shows developments in the current account and its components, while Figure 3.3b presents the pre- and post-crisis developments in various debt segments: external debt, public debt, and the two components of private debt (all expressed as a percentage of GDP). The information contained in Figures 3.3a and 3.3b provides the background to differences across countries in the build-up of external disequilibria, as well as in the debt positions of the economies on entering the crisis, which proved to be the main challenges in the subsequent adjustment processes. The evidence contained in the two figures also provides the main criteria for grouping countries in the following analysis.

Current account developments are considered first. The Central European economies (CE-5)[5] exhibit a relatively positive performance in terms of their current account developments, which did not experience any substantial deterioration before the crisis. Furthermore, the trade accounts (the light shaded bars in Figure 3.3a) confirm the relative strength of those economies in terms of their export as opposed to their import performance. A number of economies have been able to achieve positive trade balances (the Czech Republic and Slovakia) and others have come close to balance. Given this evidence, it would seem reasonable to conclude that these economies encountered no significant competitiveness problems.[6]

Much more problematic were developments in the other groups of economies. All the Baltic economies as well as Romania and Bulgaria recorded strongly deteriorating current accounts before the crisis, which were predominantly associated with deteriorating trade balances. In part, those deteriorating trade balances were a consequence of inordinately high growth rates in some of the economies prior to the crisis—to a certain extent reflecting a later start in their catching-up processes (see also Chapter 2 in this volume);

[5] The CE-5 consist of Czech Republic, Hungary, Poland, Slovakia, and Slovenia.

[6] Of course, the trade accounts per se are insufficient to reflect fully competitive strengths and weaknesses as they can, e.g., become sharply negative or positive when GDP growth exceeds or falls short of that of the main trading partners. However, as we have seen in Figure 3.1, the CEE economies have, almost without exception, experienced growth rates before the crisis substantially greater than those of their main trading partners (the more advanced Western European economies).

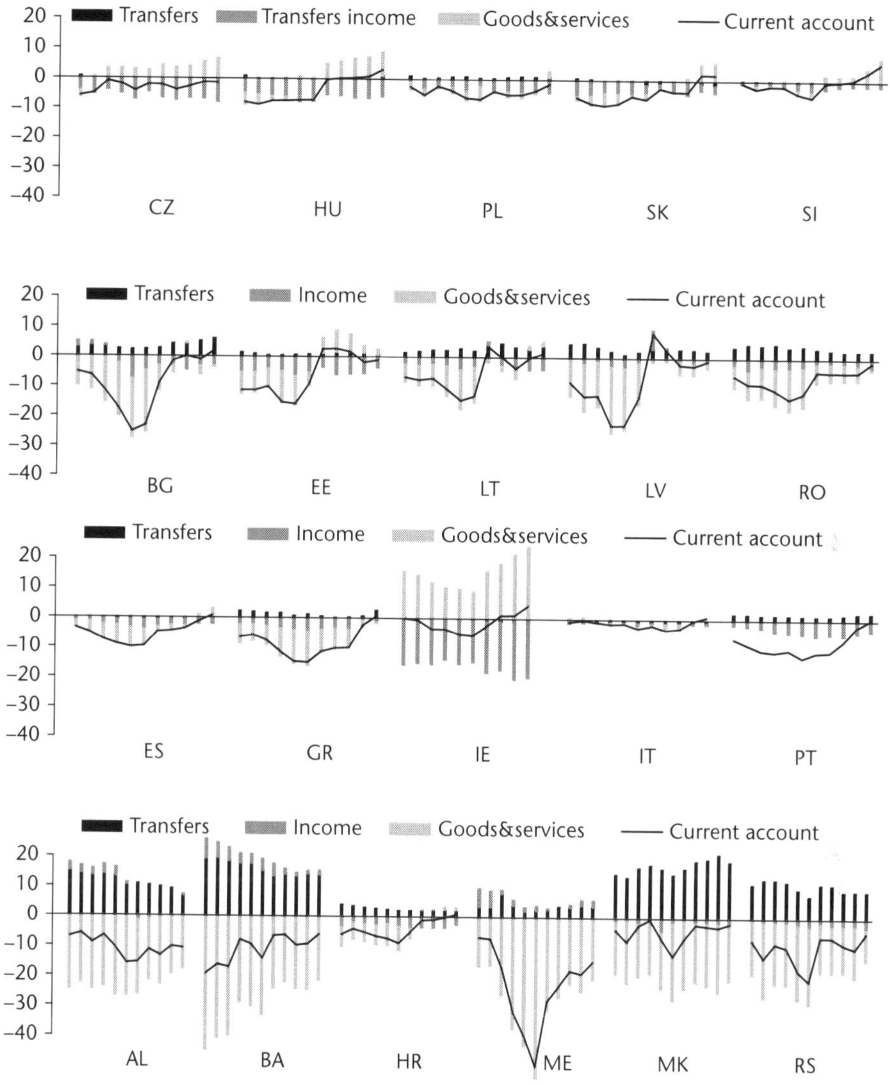

Figure 3.3a. Composition of the current account of the balance of payments, 2002–13, % of GDP

Source: wiiw Database incorporating national and Eurostat statistics. For explanation of country codes see page xix.

however, there is clear evidence here of external balances 'moving out of gear' as their trade balances in particular turned sharply negative.

As for the Southern EU economies, we can see that competitiveness problems became particularly pronounced in two countries with persistently large current account deficits, Greece and Portugal, with Spain close behind.

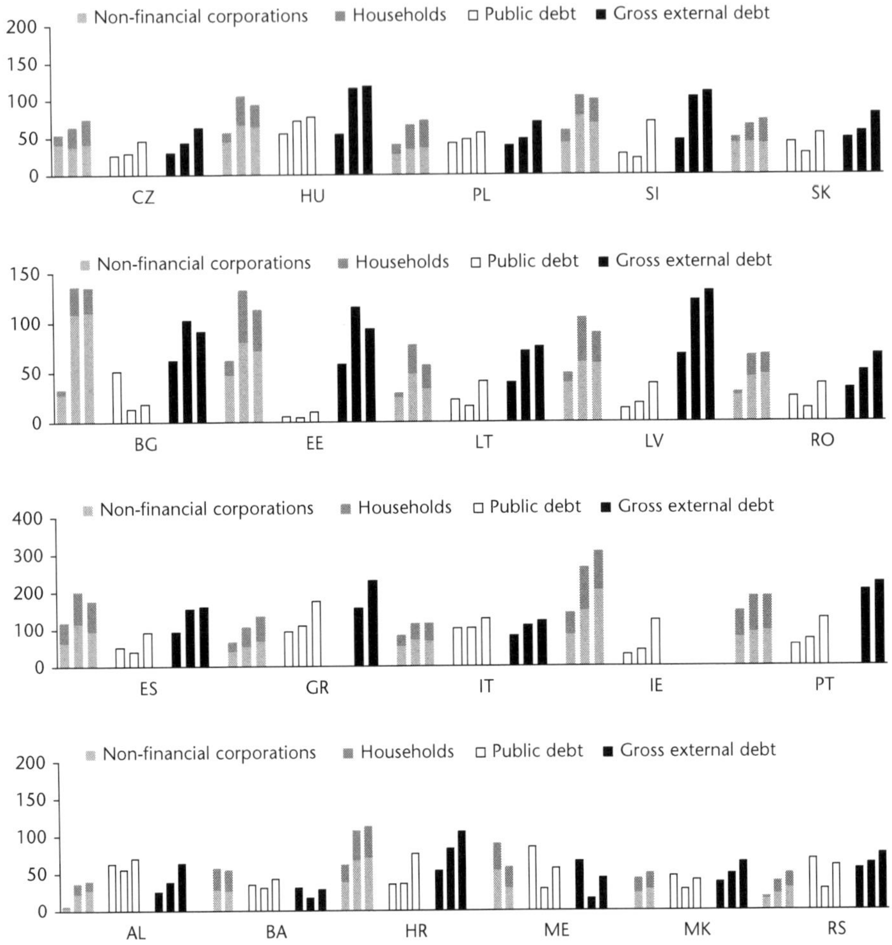

Figure 3.3b. Private and public debt, % of GDP, 2002, 2008, and 2013

Source: Eurostat, IMF, wiiw own calculations. Note the scale differences on the y-axes. For explanation of country codes see page xix.

Current account problems featured less prominently in Italy, but one must bear in mind its very low growth performance over this period (see Figure 3.1).

The remaining countries in Southeast Europe (SE-6)[7] exhibit very high trade deficits, reflecting a very small export base on which these economies can draw. Current accounts displayed a marked deterioration in the period prior to the crisis in two economies: Montenegro and Serbia. Most of the economies in the group rely on major transfers in the form of remittances from their

[7] The SE-6 consist of Albania, Bosnia and Herzegovina, Croatia, Macedonia, Montenegro, and Serbia.

nationals living and working abroad, thus partly offsetting the large trade deficit.

As is well known, current account imbalances have to be financed and the capital inflows funding these accumulate in the form of debt positions in different sectors of the economy. The accumulation of domestic and foreign debt by sectors is shown in Figure 3.3b. The information presented in the figures reveals further differences across economies.

Among the CE-5, relatively modest if any increases can be observed in the various debt levels of the Czech Republic and Poland, a somewhat higher increase of private debt in Slovakia (whereas public debt dropped as a percentage of GDP prior to the crisis), while there have been marked increases in the private-sector debt positions of Slovenia (mostly corporate debt) and Hungary. Furthermore, Hungary increased its public debt to 73 per cent of GDP in 2008, which is exceptional among the countries of Central and Eastern Europe, followed by Poland where public debt rose to 50 per cent of GDP in 2008.

The Baltic states as well as Bulgaria and Romania were characterized by the rapid development of private debt over the pre-crisis period, while public debt (as a percentage of GDP) was driven down, as it benefited from high growth rates over that period and the associated tax revenue. Among the SE-6 countries, Croatia also displayed a rapid rise in private-sector debt, while Albania was characterized by a higher level of government debt, which, however, fell (as a percentage of GDP) in the period leading up to the crisis, in common with most other economies.

The other issue that Figure 3.3b reveals is the substantial increase in public debt in quite of a few of the economies in the course of the crisis. This reflects the impact of the recession (and at times deep contractions), on the one hand, and the role that the state played in alleviating the debt position of the private sector or rescuing banks, on the other. The latter was the case in Ireland and Spain, while in the CESEE economies only a few of the banks were domestic banks, although where this was the case, severe banking crises were experienced, for example more recently in Slovenia.

Summary

In the CESEE region prior to the crisis, one group of Central European economies (the Czech Republic, Poland, and Slovakia) encountered no evident problems related to the build-up of private and public debt. In contrast, many of the other CESEE economies experienced a major build-up of private-sector debt, with only Hungary (and to a lesser extent, Poland) among the CESEE economies displaying a high public debt level relative to GDP.

The situation was not that different from that prevailing in the EU Southern economies before the crisis, with the important exceptions of Greece and

Italy, both of which maintained public debt levels of over 100 per cent of GDP prior to 2009. The other EU Southern countries (including Greece, but not Italy) experienced extremely rapid growth in private-sector debt: in other words, the (in)famous 'credit bubble' in those economies prior to the crisis.

Hence prior to the impact of the financial crisis, the starting point was major imbalances in the external accounts in a large number of low- and medium-income economies in Europe (the EU South except for Italy, and a large number of CESEE economies, although with a subgroup of CE-5 being the exception). This was accompanied by the build-up of debt, which in many economies was largely characterized by a rapid build-up of private-sector debt to very high levels, with only a small subgroup of economies (Hungary, Greece, Italy) showing high public debt levels prior to the crisis.

3.4. The Impact of the Crisis: External Accounts Adjustment

In the following, I discuss certain features of the adjustment processes that occurred in the wake of the financial and economic crisis, including both transitory features as well as the more permanent impact of the initial disequilibria on the patterns of recovery or the lack thereof.

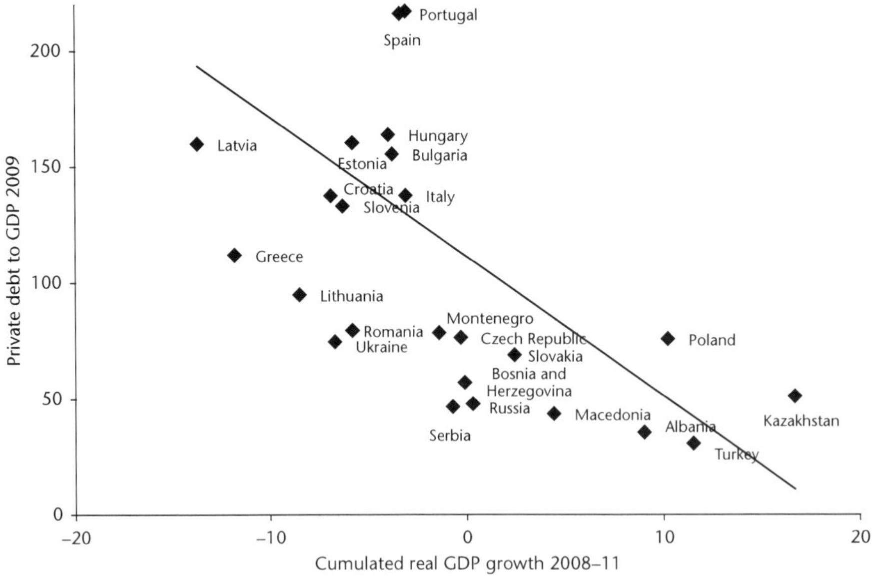

Figure 3.4a. Private-sector debt and subsequent GDP growth trajectories, 2008–11
Source: wiiw Database incorporating national and Eurostat statistics, own calculations

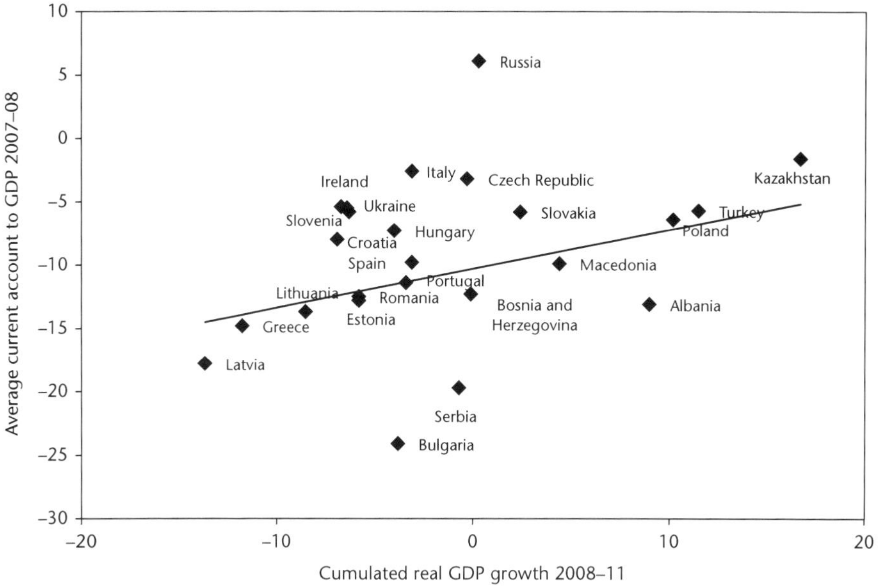

Figure 3.4b. Current account balance and subsequent GDP growth trajectories, 2008–11

Source: wiiw Database incorporating national and Eurostat statistics, own calculations

Figure 3.4a shows a clear relationship between one of the indicators of an existing disequilibrium prior to the crisis, the ratio of private-sector debt to GDP, and the subsequent growth trajectories of the different economies. A similar relationship is shown in Figure 3.4b between the pre-crisis current account deficit and subsequent growth. We can conclude that the extent of the previous build-up of private-sector debt or pre-crisis current account disequilibria had a significant negative impact on medium-term growth performance following the crisis.

The differentiation across economies is also apparent from Figure 3.5, in which both export and import growth rates are plotted over the period 2008–13, thus allowing us to distinguish between various groups of economies: (i) those in the top two quadrants which experienced relatively high export growth; (ii) those which experienced import contractions or very low import growth, but relatively weak or negative export growth (this group comprises most of the Southern EU countries, as well as quite a few of the Southeast European countries plus Slovenia and Hungary); and (iii) those with weak export growth and still substantial import growth (this group comprises the rest of the Southeast European economies).

The point of this figure is to show that there are different ways of adjusting to unsustainable external accounts, which is of importance for the future

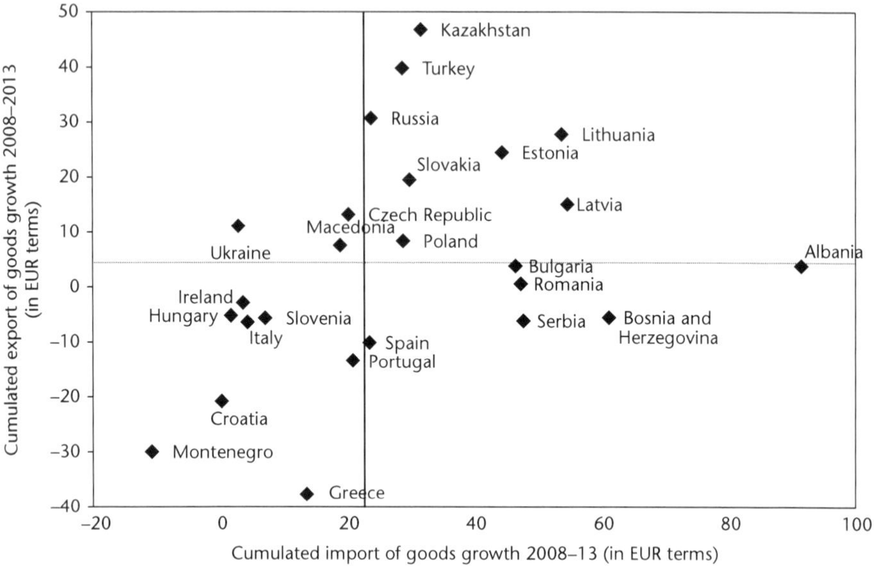

Figure 3.5. Export and import development, 2008–13

Source: wiiw Database incorporating national and Eurostat statistics, own calculations. Horizontal and vertical lines are drawn at country averages

growth prospects of 'Southern economies'. Negative external accounts can be 'adjusted' through strengthening exports or reducing imports. Figure 3.5 shows that countries followed quite different paths in this respect: some 'rebalanced' their economies through strongly improved export performance, others predominantly through contracting imports, the latter reflecting the impact of the recession (or contraction) which their economies went through. While we are not able to discuss in detail the reasons for such differentiated adjustment patterns (e.g. the roles that exchange rate regimes, wage flexibility, and monetary policy played), the impact of such differences in sectoral adjustments for future growth prospects will be discussed in the following section.

Summary

It is quite apparent that the crisis brought about a need to correct strong external imbalances and large private-sector debts that built up prior to the crisis. The extent of adjustment was directly related to the size of the previous current account disequilibria and private-sector debt build-up, and those adjustments (and their severity) imposed clear medium-term costs in terms of GDP growth. Furthermore, patterns of adjustment across economies varied greatly, with some countries relying almost exclusively (even in the medium term) on import adjustments, while others were more successful in terms of export growth.

3.5. External and Sectoral Imbalances and the 'Manufacturing Imperative' for Europe's EMEs

Sectoral Distortions and Levels of Industrial Production

In the previous section we discussed the issue of external imbalances and the need to rebalance in the face of severe reductions of capital inflows or even their reversal (i.e. 'sudden stops'; see Figure 3.3a) specifically in those economies which built up large external imbalances in the pre-crisis period. Let us now move to the need for 'rebalancing'—i.e. the strengthening of the tradable sector of those economies that experienced large current account deficits in the period prior to the crisis. Industrial production accounts for the bulk of the tradable sector in most economies and hence I am particularly interested in its development. Figure 3.6 shows levels of industrial production in relation to the pre-crisis levels. The abysmal performance of industrial production in the course of the crisis comes across clearly for the Southern EU economies, as well as in a rather wide range of Southeast European economies; industrial production levels were also strongly hit in Slovenia during the crisis. In some

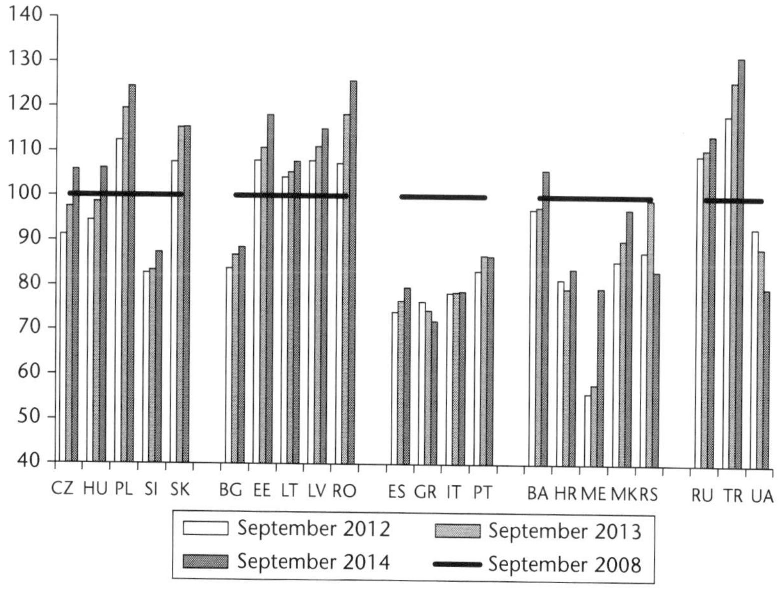

Figure 3.6. Development of industrial production, September 2008 = 100

Source: wiiw Database incorporating national and Eurostat statistics. For explanation of country codes see page xix.

countries industrial production continues to contract in 2014 (Greece, Serbia, Ukraine).

Additional information is provided in Figure 3.7, where the contributions of different sectors of the economy to overall GDP growth have been plotted.[8] Two sub-periods are shown: the pre-crisis period 2004–8 and the crisis/post-crisis period 2009–13. The following sectors are identified in this figure:[9] manufacturing (C) as the classic tradable goods sector, construction (F), a range of sectors which we categorize as 'tradable services' (TS) comprising 'other market services' (transport services, IT services, and financial and business services); non-tradable services (NTS), which comprise wholesale and retail trade (G) and real estate services amongst others; and finally non-market services (NMS).

From Figure 3.7 we can detect the following patterns:

- The manufacturing sector performed very poorly in terms of its contribution to overall GDP growth in the Southern EU countries in the pre-crisis period. GDP growth relied almost entirely on one or the other of the services sectors (TS or NTS) or on construction (F).
- In contrast, CEE economies exhibit a more balanced picture with regards to different sectors' contributions to GDP growth. Manufacturing played a particularly important role in the Czech Republic and Slovakia, while in the other CEE economies where services activities and construction dominated GDP growth, the development of manufacturing was also important.
- Manufacturing was particularly negatively affected in the Southern EU countries in the crisis period (2009–13) so that during this period there was little evidence of sectoral adjustment in favour of a strengthening of the tradable sector.

In summary, there are distinct differences between Southern EU and Eastern European economies[10] in the build-up of structural distortions before the crisis, in that the move away from manufacturing prior to the crisis was much more pronounced in the Southern EU economies and some of the CESEE economies, while the growth pattern was more balanced in Central European economies. Nonetheless, subgroups of new member states (the Baltic states, Slovenia) experienced a strong shift away from manufacturing in the period just before the crisis. As to the adjustment in the wake of the crisis, manufacturing was also very adversely affected in many countries and

[8] Contributions are calculated by multiplying the share of the selected industry by its growth rate.

[9] Based on the NACE rev. 2 classification scheme.

[10] The analysis could be carried out only for EU member countries as comparable industry-level data are not available for a wider range of CESEE economies.

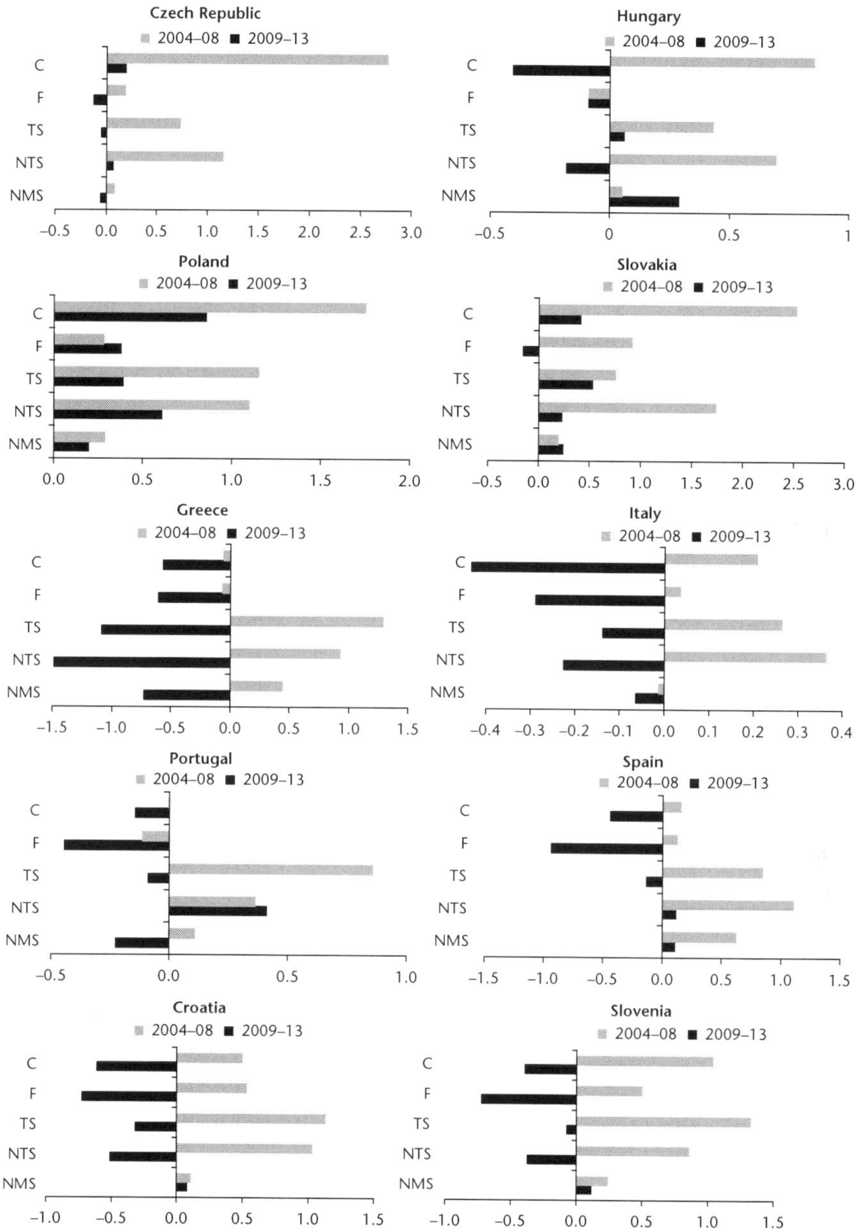

Figure 3.7. Sectoral contributions to GDP growth (from constant price data)

Notes: Based on NACE rev. 2 or NACE rev. 1 classification scheme: C (Manufacturing), F (Construction), TS (Tradable services) comprise H+J+K+M (see NACE rev.2 classification for details), NTS (Non-tradable market services) comprises G (Wholesale and retail trade) plus I+L+N+R+S+T and NMS (Non-market services) comprise O+P+Q. Contributions are calculated by multiplying the share of total GDP at current prices by real growth at preceding year prices.

Source: wiiw Database incorporating national and Eurostat statistics

hence a process of readjustment (favouring the tradable sector) has hardly started. This is particularly worrisome for those economies that entered the crisis with a weak tradable sector, chronic current account problems, and large external debts.

The 'Manufacturing Imperative' for Europe's EMEs

In the literature much has been written about the role of manufacturing at various stages of development and the move towards services. There is strong evidence that the manufacturing share in an economy increases up to a certain level of income per capita and then is expected to stabilize and subsequently to fall (e.g. Haraguchi and Rezonja, 2010; Syrquin, 2008; Rowthorn and Ramaswamy, 1997). However, other factors than just real income levels affect the share of the manufacturing sector in different economies, such as specialization in foreign trade, current account imbalances, and so on.

Figure 3.8 depicts manufacturing's share of GDP in relation to GDP per capita in 2005 (the mid-point for the pre-crisis period) for EU member countries plus Southeast Europe (SEE). The general tendency of the share of manufacturing to decline with rising real incomes in this group of economies is not confirmed in the case of the spectrum of European economies (see the weak positive slope of the regression line). What emerges clearly instead is the segmentation among the groups of economies in Europe's 'periphery': the

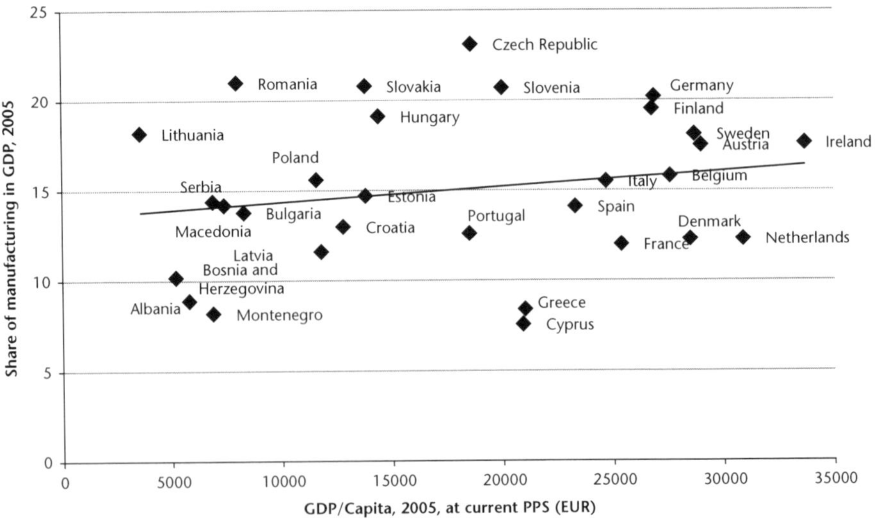

Figure 3.8. Importance of manufacturing in the economy, share of manufacturing in GDP (%), 2005, and GDP per capita at current PPS (EUR), 2005

Source: Eurostat and wiiw own calculations

group of Central European economies (Czech Republic, Slovakia, Slovenia, Hungary) and also Romania and Lithuania (in the case of the latter, this is due to its important petroleum-refining sector) which show a rather high share of manufacturing in relation to their levels of GDP per capita (i.e. they lie above the regression line), while the other Baltic states (Estonia, Latvia), Croatia, Poland, and all the Southern cohesion countries (Portugal, Greece, Spain) lie below the regression line. In addition, all SEE economies are located well below, with Albania and Montenegro having especially small manufacturing shares. Only Serbia is close to the regression line.

Among the advanced European economies we can also distinguish two groups of economies: Ireland, Germany, Austria, Finland, and Sweden, each with a very strong manufacturing sector; and France, Denmark, and the Netherlands with a low manufacturing share. Italy and Belgium lie on the regression line.

Let us now discuss the relationship between longer term trade balances and the manufacturing sector's share of GDP. Figure 3.9 shows two regression lines, one indicating the relationship between trade balances and the manufacturing share of GDP for the lower-and medium-income economies, and another showing this relationship for more advanced EU member countries. We can see that the regression line for the less advanced economies is shifted to the left compared to that for the more advanced ones; this means that lower-income economies require a higher share of manufacturing to achieve a

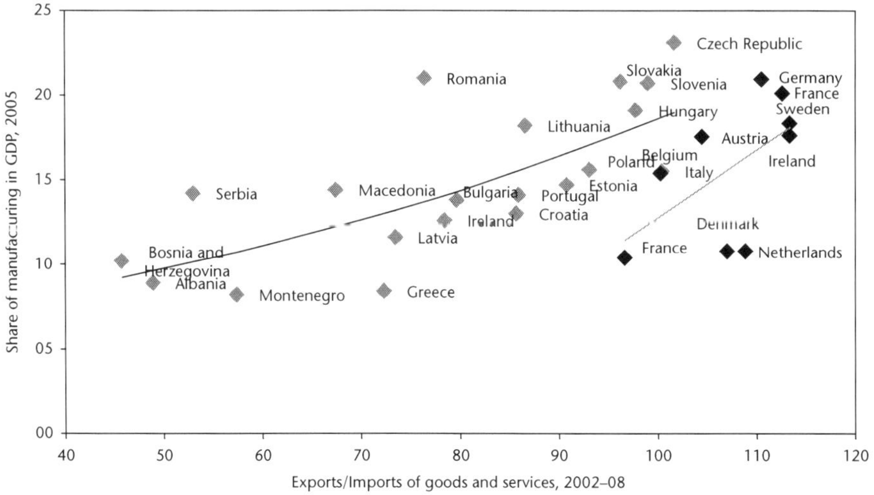

Figure 3.9. Position of manufacturing and longer term trade balances, share of manufacturing in GDP in 2005 (%), and exports/imports of goods and services, average 2002–8

Source: Eurostat and wiiw own calculations

similar balance in the trade accounts (something like a 5–7 percentage points higher share of the manufacturing sector).[11] The reason is that more advanced economies can compensate more easily for a smaller manufacturing sector by exporting tradable services than lower- and medium-income economies can. The group of economies that had particularly persistent large deficits (export/import ratios below 95 per cent) in their trade accounts in the pre-crisis period among the lower-income economies includes the Baltic economies, Bulgaria, Romania, and all the Southern cohesion economies. On the other hand, we can also see the relatively good performance of the Central European economies (Czech Republic, Hungary, Slovakia, and Slovenia).

What has been established so far is the following:

- Among the low- and medium-income economies of Europe ('Europe's South') quite distinct groups of economies can be distinguished: the group of Central European economies (Czech Republic, Hungary, Slovakia, Slovenia) as well as Bulgaria and Romania, which have a manufacturing sector accounting for a relatively high share of GDP; and the rest of the low/medium-income economies with shares which were rather below what one would expect at that level of income.
- Furthermore, in the pre-crisis period the Central European economies also boasted a healthy longer term trade balance, while Bulgaria, Romania, Croatia, the Baltics, and the Southern cohesion countries (Greece, Portugal, Spain, Cyprus) showed rather high net import positions on their external trade accounts. With the exceptions of Bulgaria, Romania, and Lithuania, these economies also had a small share of GDP in manufacturing, a factor which we have shown to be correlated with weak trade accounts.
- Among the more advanced economies there are also differences in terms of manufacturing as a share of GDP; however, this is less relevant for an explanation of longer term trade accounts, since as noted above advanced economies have more scope to specialize internationally on tradable services.
- For the group of European low-medium income economies it thus seems that a strong manufacturing sector is an important condition for longer term external equilibrium and—when the crisis hit—those European low-medium income economies with a weak manufacturing sector were therefore particularly vulnerable to the capital flow reversals (see Figure 3.3a) that accompanied the onset of the financial crisis. Given the more

[11] Figure 3.9 also shows that the exports/imports coverage is particularly low in the case of the Southeast European economies (and also Greece) in the pre-crisis period (2002–8). This indicates a non-linearity involved in this relationship, due to particular features of the region, such as the important role of remittances (in the case of the Balkan countries) and of the credit bubble (in the case of Greece and some of the other Balkan economies).

cautious attitude of international capital flows also in the years following the crisis, emerging European economies with a weak manufacturing sector can thus expect that their growth will be 'balance-of-payments constrained'.[12]

Table 3.1 provides additional evidence with regard to differentiated developments in manufacturing activity across Europe.[13] The table shows the development of shares of different country groups in European manufacturing exports (total as well as extra-EU exports) over the period 1995–2011.[14]

What we can see is that over the period 1995–2011 there was a strong *process of agglomeration of manufacturing* export capacities in a Central European manufacturing belt, which includes Germany and Austria, on the one

Table 3.1. Shares (%) and changes in shares (in p.p.) of global manufacturing value-added exports of EU Member States, 1995–2011 and in extra-EU manufacturing exports only

	Market share 1995	Share in extra-EU exports 1995	Market share 2011	Share in extra-EU exports 2011	Change in market share 1995–2011 (in p.p.)	Growth/ Decline in market share 1995–2011 (in %)	Change in share in extra-EU exports 1995–2011 (in p.p.)
Benelux	1.92	7.20	1.28	6.17	−0.64	−33.57	−1.03
Germany & Austria	9.25	34.67	7.90	38.20	−1.35	−14.61	3.53
CEEC-5	0.58	2.18	1.25	6.06	0.67	115.29	3.88
France & Italy	6.70	25.09	4.38	21.19	−2.31	−34.53	−3.90
Nordic countries	2.45	9.20	1.52	7.36	−0.93	−37.99	−1.84
Baltic countries	0.05	0.17	0.09	0.42	0.04	90.95	0.25
Bulgaria & Romania	0.14	0.51	0.22	1.07	0.09	64.00	0.56
United Kingdom	3.99	14.95	2.16	10.44	−1.83	−45.84	−4.50
Ireland	0.39	1.47	0.63	3.03	0.24	60.37	1.57
Southern Europe	1.22	4.58	1.25	6.06	0.03	2.54	1.48
EU-27	**26.69**	**100.00**	**20.68**	**100.00**	**−6.00**	**−22.50**	**0.00**

Notes: Industry classification based on NACE Rev. 1.1. CEEC-5 = Poland, Czech Republic, Slovakia, Slovenia, Hungary; Nordic countries = Denmark, Finland, Sweden; Southern Europe = Greece, Spain, Portugal, Cyprus, Malta. Global market shares in value added exports and changes thereof exclude intra-EU value added exports
This table is from Stoellinger et al., 2013, 13, table 2.3
Source: WIOD, wiiw calculations

[12] See the literature associated with the writings of A. J. Thirlwall (e.g. Thirlwall, 1979, 2011).

[13] We adopt here a more detailed grouping of countries in order to show differentiation both among the more advanced and the lower-income countries of Europe.

[14] We make use here of recent research results calculating *trade in value-added* figures derived from the WIOD database; see Stoellinger et al., 2013.

hand, and Central Eastern Europe (the CE-5), on the other. Such a tendency, assuming it results from structural processes well explored in the economic geography literature, such as locational economies of scale and the importance of geographic proximity for trade and logistic costs, could make it difficult for economies that are not currently part of the Central European 'manufacturing core' to build up sufficient export capacities to escape from the trap of balance-of-payments constrained growth in the future.

The tendencies apparent in Table 3.1 are given further support by Figure 3.10, which shows the relationship between two variables: the differential growth between manufacturing and GDP (both at constant prices) over the period 2002 to 2013 and the share of manufacturing in 2005 (i.e. some years before the outbreak of the crisis). What we observe is a positive relationship between these two variables, which corresponds to a picture of 'divergence': in other words, countries on Europe's periphery which already had a low manufacturing share experienced lower growth rates in manufacturing than in the economy as a whole, while countries with a strong manufacturing sector (the Central European economies) experienced a positive growth differential of manufacturing relative to the economy as a whole. Hence the relative weaknesses and strengths of different countries with regard to manufacturing became *more pronounced* over that period (see also Havlik, 2014).

The importance of manufacturing for overall exports growth is also shown in Figure 3.11, which plots differential growth in manufacturing compared with the economy as a whole against export growth (total exports). Again, we

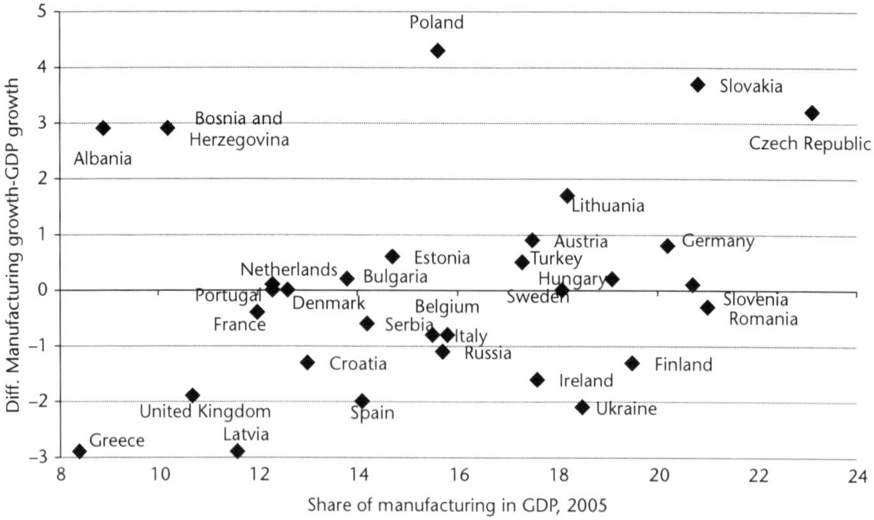

Figure 3.10. Differential growth manufacturing—total economy, 2002–13 (p.a.) and share of manufacturing in GDP, 2005

Source: Eurostat and wiiw own calculations. Growth rates are calculated using constant price data

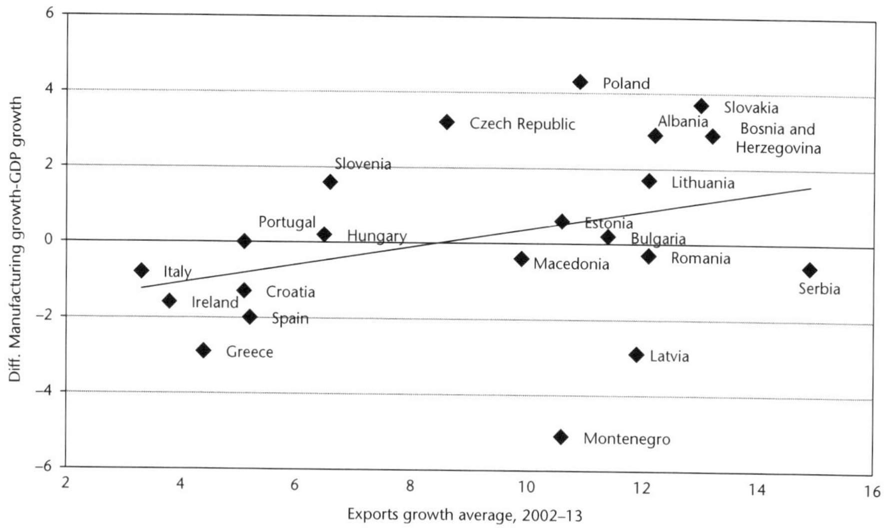

Figure 3.11. Differential growth manufacturing—total economy, 2002–13 and (total) export growth

Source: Eurostat and wiiw own calculations. Growth figures refer to average growth rates per annum; export growth rates were calculated from current EUR price data

observe a picture of 'divergence', in that economies with low export growth also experienced relatively low growth in manufacturing compared with the economy as a whole, which means that in these economies the economic structure shifted further away from manufacturing. Such a trend is particularly worrying for low- and medium-income economies in which manufacturing constitutes the most important part of the tradable sector and thus the most crucial for potential export growth.

Summary

In this section we have seen that the manufacturing sector is particularly important for lower- and medium-income countries (as compared to more advanced economies which can rely on a stronger contribution by the tradable services sector) to make sure an economy does not suffer from chronic longer term trade imbalances. This provides the basis for a 'manufacturing imperative' for low- and medium-income countries in order to avoid a 'trade-balance constraint' on growth and catching-up in a post-crisis world in which sustained current account disequilibria can no longer be financed as easily as prior to the crisis.

I have also presented evidence for 'structural divergence' in that countries with a weak manufacturing sector were moving further away from manufacturing. This was not the case for the Central European countries which saw a strengthening—with the exception of Slovenia—in the position of the manufacturing sector. The overall economic geography of Europe is characterized by manufacturing capacities being concentrated in a 'Central European core' (Germany, Austria, and Switzerland, on the one hand, and CEE-5, on the other). Given the importance of manufacturing in avoiding longer term structural imbalances on the current account, especially for lower- and medium-income economies, this creates a serious problem for large segments of 'Europe's periphery' from a longer term development perspective.

3.6. Implications of 'Divergence': Labour Markets and Political Economy Considerations

The implications of the trends examined in this chapter (longer term imbalances in trade structures, agglomeration tendencies in European manufacturing, reinforcement of weaknesses in the tradable sectors for a range of European EMEs during the crisis) clearly should not be taken lightly. Labour market effects of the recent crisis have been severe (see Figures 3.12 and 3.13). Overall unemployment rates have, in the best performing countries (EU North and CEE-Central), returned to the relatively high levels of the early 2000s. In a second group of EU members (Bulgaria, Romania, Croatia), unemployment rates now exceed those of the early 2000s and are still rising. Finally, unemployment rates in the EU South (Greece, Portugal, Spain) have now

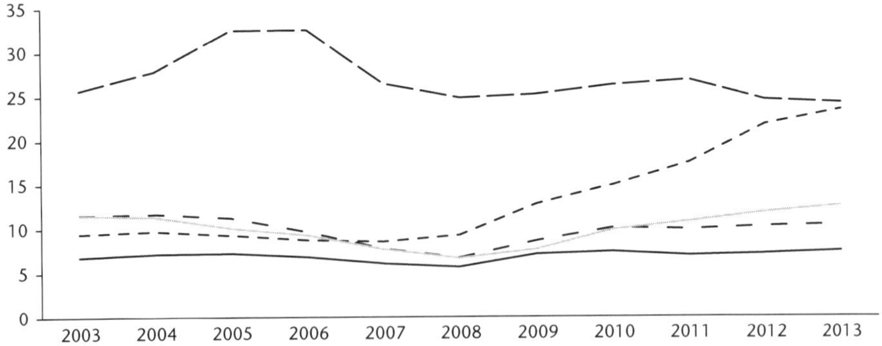

Figure 3.12. Unemployment rate in %, 2003–13

Remark: EU North: Belgium, Denmark, Germany, France, Netherlands, Austria, Finland, Sweden, UK. EU South: Greece, Spain, Portugal. CEE-Central: Czech Republic, Hungary, Poland, Slovenia, Slovakia. CEE-3: Bulgaria, Croatia, Romania. WB-6: Albania, Bosnia and Herzegovina, Kosovo, Macedonia, Montenegro, Serbia. Average over available countries.
Source: Eurostat and national statistics, wiiw calculation

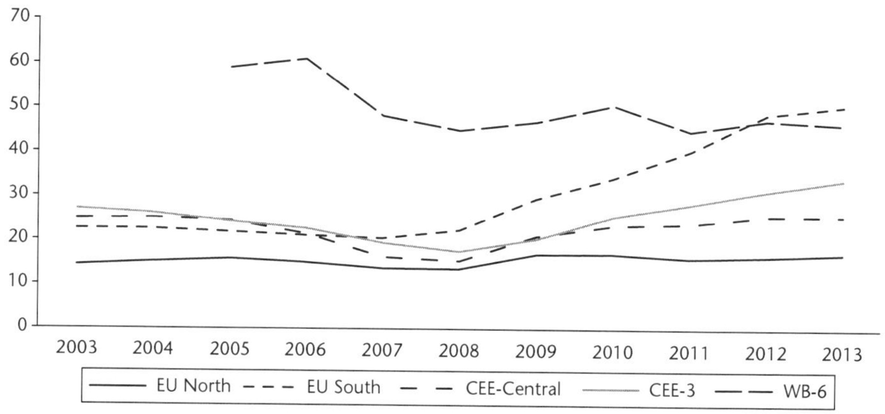

Figure 3.13. Youth unemployment rate in %, 2003–13

Remark: EU North: Belgium, Denmark, Germany, France, Netherlands, Austria, Finland, Sweden, UK. EU South: Greece, Spain, Portugal. CEE-Central: Czech Republic, Hungary, Poland, Slovenia, Slovakia. CEE-3: Bulgaria, Croatia, Romania. WB-6: Albania, Bosnia and Herzegovina, Kosovo, Macedonia, Montenegro, Serbia. Average over available countries

Source: Eurostat and national statistics, wiiw calculation

reached (and for youth unemployment exceeded) those in the Western Balkans, a region ravaged by the Yugoslav wars and which continues to be characterized by very poor labour market indicators (unemployment rates, share of informal employment, etc.)

The youth unemployment figures are particularly alarming in that a clear 'centre–periphery' differentiation is visible. Two groups of Europe's 'South' (EU South and the Western Balkans) exhibit youth unemployment rates of about 50 per cent and the CEE-3 group (Bulgaria, Romania, Croatia) rates of over 30 per cent. Recent developments in the Ukraine and the ever precarious situation in the Caucasus region add to the picture of highly uneven labour market conditions in Europe. One of the developmental implications is a high migration rate, with the younger and more skilled labour force in particular leaving their countries; this, in turn, has a detrimental impact on demography and the available skill mix in these countries. Furthermore, research has shown that the development of a 'remittance dependency' culture has adverse consequences for labour supply (activity rates), for the educational performance of children and—at the macroeconomic level—for real exchange rate developments, all of which further reduces competitiveness (see e.g. Ratha, 2013, and references therein).

The labour market developments depicted here come on top of the direct effects of the policy of fiscal consolidation and its distributive impact on different groups of society. These have significant social effects, particularly in countries 'falling behind' but also in better performing European countries as well. Furthermore, in an integrated Europe, attempts to alleviate some of the 'divergence' problems through debt resolution and transfer mechanisms generate tensions between countries regarding 'burden sharing' that reduce the willingness to further the integration process just at a time when further 'deepening' and policy coordination (especially regarding bank restructuring and fiscal policy) may be required to manage the crisis (see e.g. Corsetti et al., 2015).

The other effect is in an erosion of democracy. This can take a number of forms that are well known from history. The problem is compounded in the EU context, given the additional issue of national sovereignty and the perception that decisions are being imposed from above. These social and political developments may seriously challenge the set-up of the EU and Eurozone policy mix. In all likelihood, a prolonged period of stagnation or low growth will lead to significant social and political changes in the divergent countries and, in turn, to instability in the EU and the Eurozone.

3.7. Summary and Policy Conclusions

This chapter has argued that trust in the 'European convergence model' has been severely dented by developments both pre-crisis and during the crisis itself, and this constitutes a serious challenge for the growth and integration project in Europe as a whole.

At the root of this challenge lie certain 'structural current account problems'[15] in a range of European low- and medium-income economies (in what I termed 'Europe's South') and these in turn have a number of causes:

- In the case of the Eastern European transition economies, one cause relates to the dramatic process of 'deindustrialization' that characterized the first phase of transition and from which a significant number of Eastern European economies have still to recover.
- Processes involving 'structural distortions' which took place in the decade prior to the crisis, when in a number of Southern EU and Eastern European economies there was a sustained bias in favour of the 'non-tradable' sector and hence a relative shrinking of export capacity took place. Furthermore, in the pre-crisis period a cumulative divergence in such processes was observed across European economies: countries with a stronger share of tradable sectors experienced a further strengthening and those with a weaker share a further weakening.

[15] By 'structural current account' problems, we mean that inadequate export capacities do not allow an economy to grow at a reasonable rate (in the case of catching-up economies at a rate higher than that of advanced economies) without encountering an unsustainable build-up of debt and eventually financing problems.

- A number of factors account for such 'structural divergence' processes. (i) Vigorous financial market integration encouraged strong credit booms in the private sector and led to investment in the non-tradable sector; the policy choices of economies where historically weak monetary authorities opted for fixed or pegged exchange rate regimes played a role in this. (ii) There are signs that manufacturing activity in Europe is becoming geographically more concentrated in a 'core area' and it seems difficult for other (peripheral) economies and regions to enter such evolving cross-border production networks; this is a particular problem for low- and medium-income economies which are much more dependent on manufacturing to provide the bulk of exports. (iii) Chapter 2 in this book has furthermore pointed out that certain economies were particularly affected by strong Chinese and other global EME imports into Europe, given their historical specialization structures in international trade.
- The fear in the wake of the crisis is that the unbalanced nature of the development of tradable sectors (and thus of export capacities) cannot easily be reversed: the crisis itself has hit the manufacturing sectors in many of the weak economies, and the expected period of reduced FDI inflows in the post-crisis era will make it more difficult for the (structurally) weak economies to mimic the favourable processes of technology transfer that characterized the successful pre-crisis developments in Central European economies.

What are the Policy Implications from this Looming 'South–North Divide' in Europe?

If Europe is to return to a 'convergence phase', in other words with low- and medium-income economies again growing at a faster rate than the more advanced economies, it is of utmost importance that policy instruments are intensively directed to make up for the smaller role that FDI is likely to play in the post-crisis environment and also to counteract to some extent the underlying dynamic of further agglomeration of industrial activity in Europe. Using such instruments should thereby encourage a wider range of countries to participate in cross-European industrial networks.

I therefore welcome the emergence of *new forms of industrial policy* at national and European levels to support the build-up of industrial capacity in Europe's periphery. There is insufficient space here to elaborate on the characteristics of effective forms of 'new' industrial policies[16] that would avoid the pitfalls of some of the 'older forms', but in short they should be

[16] See Ch. 9 by Mazzucato and Perez for a more detailed discussion of such policies.

strongly focused on the tradable sector, support new entrants (much of 'old industrial policy' favoured incumbents), emphasize new activities and products thus contributing to a process of industrial diversification, be aware of the importance of and facilitating the linking up with international production networks, and build up contextual conditions (education, research facilities, infrastructure) which support such an orientation. There are obvious constraints on the use of industrial policy instruments given the EU's competition policy and the rules of the Single Market. However, an industrial policy directed towards new entrants, new activities, and new products, and providing good infrastructure should be more palatable to EU competition authorities. In addition, the obvious potential threat of an entrenched 'North–South Divide' for the European integration process should motivate initiatives to find a new balance between competition and industrial policies, a balance that allows the asymmetric use of industrial policy instruments in favour of weaker countries and regions.

A shift of policy is all the more necessary as political developments in the course of the crisis have shown the limits of Europe developing into a 'Transfer Union'. A sustained 'convergence' process requires that every country has an export sector of above the critical size needed to escape a structural current account disequilibrium. This has to be a top priority of national and European policy-making at the present time.

Acknowledgements

Scientific Director, The Vienna Institute for International Economic Studies (wiiw); Rahlgasse 3; Vienna A-1060; Austria; landesmann@wiiw.ac.at;www.wiiw.ac.at. The chapter draws on joint work, particularly with Vladimir Gligorov, Mario Holzner, and Doris Hanzl; thanks go also to the wiiw statisticians, particularly to Beate Muck and Galina Vasaros.

References

Becker, T., Daianu, D., Darvas, Z., Gligorov, V., Landesmann, M., Petrovic, P. J., Pisani-Ferry, J., Rosati, D., Sapir, A., and Weder di Mauro, B. (2010). *Whither Growth in Central and Eastern Europe? Policy Lessons for an Integrated Europe*, Brussels and Vienna: Bruegel Blueprint, 11; Bruegel and wiiw.

Corsetti, G., Feld, L. P., Garicano, L., Lane, P. R., Reichlin, L., Rey, H., Vayanos, D., and Weder di Mauro, B. (2015). *A New Start for the Eurozone: Dealing with Debt*, Monitoring the Eurozone, 1, London: CEPR Press.

Darvas, Z. (2012). 'Intra-Euro Rebalancing is Inevitable, But Insufficient', *Bruegel Policy Contribution*, 2012(15), Aug.

De Grauwe, P. (2012). *In Search of Symmetry in the Eurozone*, Brussels: CEPS Policy Brief, 268.

European Commission (2013). *Quarterly Report of the Euro Area*, 12(4).

Gerschenkron, A. (1962). *Economic Backwardness in Historical Perspective: A Book of Essays*, Cambridge, Mass.: Belknap Press of Harvard University Press.

Haraguchi, N., and Rezonja, G. (2010). *Patterns of Manufacturing Development Revisited*, Vienna: UNIDO, Research and Statistics Branch, Working Paper 22/2009.

Havlik, P. (2014). *Patterns of Structural Change in the New EU Member States*, wiiw Research Report, 394, Vienna: Vienna Institute for International Economic Studies (wiiw).

Hunya, G., and Schwarzhappel, M. (2013). *Growth Engine Stutters*; wiiw FDI Report 2013–06, Vienna: Vienna Institute for International Economic Studies.

IMF (2014). *World Economic Outlook: Update*, Washington, DC: IMF.

Ratha, D. (2013). *The Impact of Remittances on Growth and Poverty Reduction*, Washington, DC: Migration Policy Brief, 8, Migration Policy Institute.

Rowthorn, R., and Ramaswamy, R. (1997). 'Deindustrialization: Causes and Implications', *Staff Studies for the World Economic Outlook*, Washington, DC: International Monetary Fund, 61–77.

Stoellinger, R., Landesmann, M., Stehrer, R., Poeschl, J., Holzner, M., and Foster-McGregor, N. (2013). *A 'Manufacturing Imperative' in the EU: Europe's Position in Global Manufacturing and the Role of Industrial Policy*, Brussels: DG Enterprise, *Background Report to the EC Competitiveness Report 2013*.

Syrquin, M. (2008). 'Structural Change and Development', in A. K. Dutt and Jaime Ros (eds), *International Handbook of Development Economics*, Cheltenham: Edward Elgar, 48–67.

Thirlwall, A. P. (1979). 'The Balance of Payments Constraint as an Explanation of International Growth Rate Differences', *Banca Nazionale del Lavoro Quarterly Review*, 32(128): 45–53.

Thirlwall, A. P. (2011). 'Balance of Payments Constrained Growth Models: History and Overview', *PSL Quarterly Review, Economia civile*, 64(259): 307–51.

Part II

The Transition to a Sustainable Economy

4

The Arduous Transition to Low-Carbon Energy

A Multi-Level Analysis of Renewable Electricity Niches and Resilient Regimes

Frank W. Geels

4.1. Introduction

Climate change is a complex problem, related to different kinds and sources of greenhouse gas (GHG) emissions. According to the IPCC (2007), energy supply is the largest-contributing sector (26 per cent), but industry (19 per cent), land-use and forestry (17 per cent), agriculture (14 per cent), and transportation (13 per cent) also make substantial contributions. This sectoral complexity means that there is no single solution to the climate change problem. Rather, efforts to address climate change are likely to vary between sectors (energy, transport, buildings, agro-food, industry, waste) and countries. This chapter focuses on low-carbon transitions in the energy supply sector, primarily electric power. These transitions are commonly described as changes in socio-technical systems (Elzen et al., 2004; Geels, 2005), which refer to shifts from one energy, transport, housing, or agro-food system to another. These transitions entail not only changes in technology, but also in markets, regulations, culture, infrastructure, user practice, and science. With regard to climate change, the promise of transitions is that shifts to new systems may be able to deliver the large cuts in GHG emissions (up to 80 per cent by 2050) that are needed (Elzen et al., 2004).

The chapter focuses on energy supply for two reasons. First, this sector is responsible for the highest relative contribution to GHG emissions, which makes it intrinsically important for addressing climate change. Second, this is one of the few sectors in which some progress has been made in the last decade. In its evaluation of worldwide developments, the IEA (2013b, 7), for instance, remarks that: 'renewable energy and emerging country efforts are lights in the dark as progress on clean energy remains far below a 2-degree pathway'. Although renewable energy shows some progress, the IEA (2013b) characterizes developments in other energy supply-related areas either as 'not on track' (nuclear power, coal-fired power, carbon capture and storage) or as 'improvement, but more effort needed' (gas-fired power, smart grids).[1] This tension between positive and negative developments underpins the analytical contribution of this chapter, namely that studies of low-carbon transitions should address both the emergence of new 'green' options (such as renewable energy) and the defence and resistance from existing systems (such as coal, gas, and nuclear). If tensions between 'new' and 'old' are substantial in the energy domain, which is often seen as having made most progress, then low-carbon transitions in other domains can be expected to be even more difficult. These tensions also suggest that the multi-level perspective (MLP), discussed in this chapter, is a useful framework to analyse socio-technical transitions, because it explicitly addresses multi-dimensional struggles between (green) niche-innovations and existing regimes.

Against this background, the chapter has two aims. The first aim is to provide a descriptive overview of positive and negative developments with regard to low-carbon transitions in electricity generation, both globally and in Europe, which is the focus of this book. The second aim is to highlight and analyse the resilience (through innovation) and resistance (through political strategies) of incumbent actors in electricity generation. This focus should enable us to derive a deeper understanding of why low-carbon transitions are arduous, facing many uphill struggles. Furthermore, while many scholars have already studied the emergence of new options, only a few have investigated the stability and destabilization of existing regimes (see Turnheim and Geels, 2013, for an exception). This regime analysis focuses on coal, gas, and nuclear in UK power generation. The reason for focusing on a specific country is that this enables a more detailed study of the beliefs and strategies of concrete actors. The UK has been chosen because its Climate Change Act (2008) is one of the most ambitious policy frameworks, with explicit long-term goals (80 per cent reduction of GHG emissions by 2050, compared to 1990) and intermediate goals (e.g. cutting GHG emissions by 35 per cent by 2022) enshrined in law. So, if the commitment and resistance of incumbent actors remains strong in this jurisdiction, similar patterns may well occur in other countries (although country specificities in terms of public concerns,

[1] The IEA also characterizes biofuels and buildings as 'not on track', and industry and fuel economy of cars as 'improvement, but more effort needed'. The only other category that is labelled as 'on track' is electric and hybrid-electric vehicles.

political culture, industry structure, and resource endowments are nevertheless important).

The chapter is structured as follows. Section 4.2 briefly introduces the multi-level perspective which informs the analyses. Although the MLP can be used to provide detailed analyses of recursive interactions between agency and structures (rules and institutions), this chapter uses it more superficially to characterize global green niche-innovations as struggling against resilient regimes (coal, gas, nuclear). Section 4.3 describes global and European developments with regard to low-carbon electricity transitions. For reasons of space, this description aims to make sense of the main developments, which means that many important details are not addressed. Section 4.4 focuses on the resilience of existing regimes in UK electricity generation, focusing on incumbent actors such as utilities and national policy-makers. The main conclusions are summarized in section 4.5.

4.2. Multi-Level Perspective on Transitions

The multi-level perspective combines insights from evolutionary economics, sociology of technology, and neo-institutional theory (Geels, 2002; 2012; Smith et al., 2005; Geels and Schot, 2007). The MLP understands transitions as arising from the interplay between multi-dimensional developments at three analytical levels: niches (the locus for radical innovations), socio-technical regimes (the locus of established practices and associated rules), and an exogenous socio-technical landscape. Niche-innovations tend to be pioneered by entrepreneurs or relative outsiders. Incumbent actors may subsequently also engage in niche-activities as part of their hedging or diversification strategies. Examples are car companies that now also engage in developing various electric vehicles, or utilities operating wind parks. Initially, however, niche-innovations tend to struggle against existing socio-technical systems, which are reproduced and incrementally changed by incumbent actors (firms, policy-makers, special-interest groups, mainstream consumers, wider publics). The interpretations and activities of these actors are shaped by socio-technical regimes, which are deep-structural rules and institutions (belief systems, capabilities, formal regulations, normative role expectations). Innovation in existing regimes is mostly incremental because of lock-in mechanisms and path dependence (Unruh, 2000).

The MLP conceptualizes the dynamics of transitions as emerging from alignments between processes at the different levels (Figure 4.1): (a) niche-innovations build up internal momentum (because of increasing investments, price/performance improvements, support from powerful actors); (b) changes at the landscape level (macro-political ideologies, societal values and

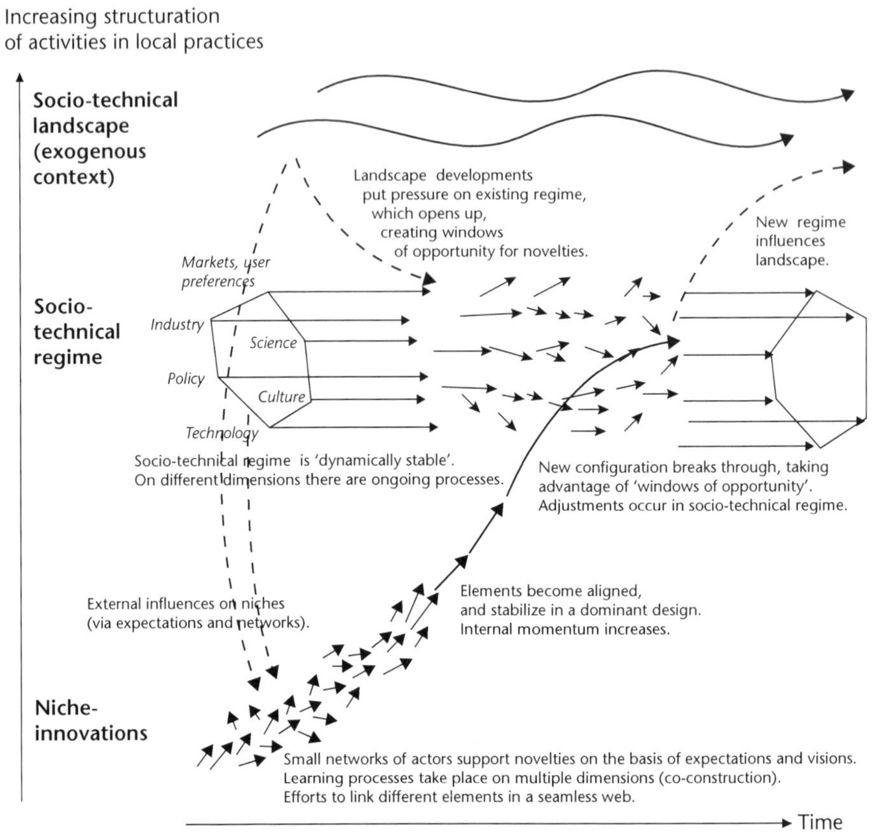

Figure 4.1. Multi-level perspective on transitions
Source: Adapted from Geels, 2002, 1263

concerns, macroeconomic trends, wars, natural environment) create pressure on the regime; and (c) destabilization of the regime creates windows of opportunity for niche-innovations. There is no guarantee that transitions will succeed: niche-innovations may fail to build up sufficient momentum or suffer setbacks; tensions in existing regimes may remain small so that 'windows of opportunity' for niche-innovations do not materialize.

Conceptually the MLP has the following characteristics:

1. Co-evolution. Transitions are not just about techno-economic processes, but also about social, cultural, and political struggles. The co-evolution between these processes requires a multi-dimensional analysis.
2. Interactions between new *and* existing configurations. Studies of transitions should not just look at radical niche-innovations (e.g. wind turbines, solar-PV), but also at existing regimes.

3. Actor-based approach. The MLP focuses on specific systems (e.g. transport, energy, agro-food) and the beliefs, interests, strategies, resources, and actions of associated actors. In this sense, the MLP differs from long-wave theory (Freeman and Perez, 1988), which focuses on entire economies and more impersonal techno-economic forces and socio-institutional frameworks.

4.3. A Global Overview of Transitions towards Low-Carbon Electricity Systems

The debate on climate change and low-carbon transitions is confusing, because of conflicting trends and reports. On the one hand, we hear much about new wind parks, transition towns, electric cars, and renewable energy. On the other hand, new coal-fired power plants are being built, sales of petrol-driven cars are booming in BRIC countries (including SUVs), and GHG emissions are continuing to rise, with CO_2 concentrations crossing the 400 parts per million (ppm) point in May 2013. In terms of the MLP, these conflicting developments can, at least partly, be understood as being linked to the emergence of green niche-innovations and the continuation of existing regimes (lock-in). This section discusses some of the main 'positive' and 'negative' developments, which are to some extent represented in Figure 4.2. This figure shows the general rise of global GHG emissions, with a minor decrease in 2009, followed by accelerated emissions increases thereafter. However, the figure also shows that emissions have decreased in recent years in the United States and Europe, partly because of macroeconomic factors (recession and relocation of manufacturing industries to other countries), partly because of energy-specific reasons (a shift from coal to gas in the US and increasing penetration of renewable electricity in Europe). In China, India, and other non-OECD countries, however, emissions have been rising very fast, often from a relatively low base, because of industrialization, rising car-transport, and raising millions of people out of poverty (who want homes, TVs, refrigerators etc.).

Table 4.1 provides a more detailed overview of the electricity mix of various countries. It shows that fossil fuels (coal, gas) play a dominant role, particularly in China and India, but also in the UK and US. Nuclear power is substantial in Europe and the US, but not yet very developed in BRIC countries (except for Russia). Interestingly, global electricity from renewable sources is also quite substantial, about 20.5 per cent in 2011 and 21.7 per cent in 2012 (REN-21, 2013). The bulk of renewable power was generated by hydro-electric power, which is often labelled as an 'old' renewable, because it is a mature technology and because most suitable sites are already in use, which limits

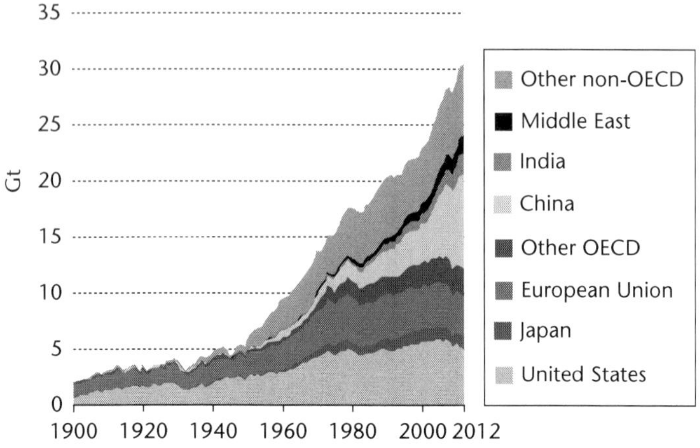

Figure 4.2. Energy-related CO_2 emissions by country
Source: IEA, 2013a, 30

Table 4.1. Relative composition (%) of electricity production in 2011

	World	EU	Germany	UK	US	China	India	Brazil
Fossil fuels	67.8	50.3	60.2	70.4	68.1	81.5	81.2	8.7
Nuclear	11.7	27.6	17.5	18.9	18.9	1.8	3.1	3.1
Total renewables	20.5	22.1	22.3	10.7	13.0	16.7	15.7	88.2
'Old' renewables (hydro-power)	16.3	10.4	4.0	2.4	8.1	14.7	13.1	82.6
'New' renewables	4.2	11.7	18.3	8.3	4.9	2.0	2.6	5.6
- wind	2.1	5.5	8.0	4.2	2.8	1.9	2.4	0.4
- biomass and waste	1.5	4.6	7.1	4.0	1.6	0.1	0.2	5.2
- solar	0.3	1.4	3.2	0.1	0.1	0.04	0.04	0.007
- marine energy	–	0.01	–	–	0	0	–	–
- geo thermal	0.3	0.2	0.003	–	0.4	0.004	–	–

Source: Constructed from data in Observatoire, 2012

possible expansion.[2] 'New' renewables (wind, solar, biomass and waste-to-power, geothermal, marine technologies) provide much less power, but do experience substantial growth rates (see later in the chapter). Most 'new' renewables are situated in Europe, as a result of cumulative investments over the past decade.

[2] There are also heated debates about the sustainability of hydro-electric power.

Positive Developments in Low-Carbon Electricity Transitions

The last decade saw strong worldwide growth in installed capacity of renewable electricity options.[3] As is visible in Figure 4.3, various green niche-innovations have grown substantially, wind power about 30 per cent annually for several years; solar-PV about 60 per cent per year in the last two to three years, reaching the 100GW threshold in 2012.

Since 2004, Europe has cumulatively invested most in the installation of 'new' renewable capacity (Table 4.2). In 2012, however, European investments decreased by 29 per cent and in 2013 by another 44 per cent, because of austerity pressures and cuts in renewable energy incentives. Investments in China, in contrast, continued to increase in 2012, making it the country with the single largest renewable energy investments (even though investments decreased somewhat in 2013). This expansion of renewables, and energy efficiency improvements, resulted in a decrease in Chinese CO_2 emissions per unit of power generation from 885 gCO_2/kWh in 2003 to 735 gCO_2/kWh in 2012 (IEA, 2013a). Table 4.2 also shows that investments in US (and Brazil) have experienced many ups and downs, which creates an uncertain investment climate. The 34 per cent decrease in 2012 and 10 per cent decrease in 2013 were related to the expiry of renewable energy subsidy programmes and

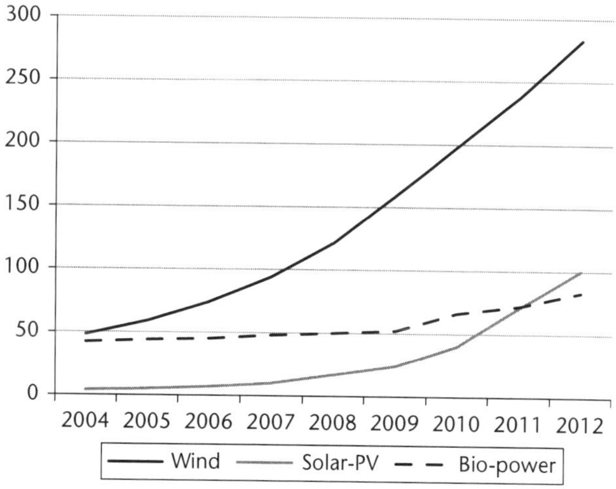

Figure 4.3. Worldwide installed capacity of wind, solar-PV, and bio-power (wood, waste) in GW

Source: Constructed from data in REN-21, 2013

[3] It should be noted, however, that capacity is not the same as actual power generation. Because wind and solar-PV are intermittent sources, actual generation is much less than capacity. Zehner (2012), for instance, suggests that in the US, wind operates at 23% of capacity and solar-PV at 16%. In contrast, US capacity factors for nuclear and coal are 91% and 73%, respectively.

Table 4.2. New investment in renewable energy (excluding large hydro) by geography in $bn

	2004	2005	2006	2007	2008	2009	2010	2011	2012	2013	2012–2013 growth
Europe	19.7	29.4	39.1	61.8	73.4	75.3	102.4	114.8	86.4	48.4	−44%
China	2.4	5.8	10.1	15.8	24.9	37.1	36.7	51.9	59.6	56.3	−6%
US	5.5	11.7	28.2	33.6	35.9	23.5	34.7	53.4	39.7	35.8	−10%
ASOC (excl. China & India)	6.8	8.2	9.0	10.9	11.4	12.9	20.7	25.3	29.5	43.3	47%
Middle East and Africa	0.5	0.5	0.9	1.6	2.3	1.4	4.3	3.2	10.4	9.0	−14%
AMER (excl. US & Brazil)	1.4	3.3	3.2	4.9	5.8	6.1	11.5	8.7	9.9	12.4	26%
India	2.5	2.9	4.4	6.3	5.4	4.2	8.7	12.6	7.2	6.1	−15%
Brazil	0.6	2.6	4.6	11.0	12.2	7.8	7.7	9.7	6.8	3.1	−54%
Total	39.5	64.5	99.6	145.9	171.2	168.4	226.7	279.4	249.5	214.4	−14%

Source: Constructed from data in BNEF, 2014

the ongoing shale gas revolution, which competes with renewables. Table 4.2 also shows increasing investment momentum in the Americas (excluding the US and Brazil) and in the Middle East and Africa (although investments decreased somewhat in 2013). This points to a wider trend, namely a relative shift in investment focus from developed to developing countries (which the IEA (2013b) also signalled as one of the positive trends): total investment in developing countries rose 19 per cent in 2012 to $112 billion, while investment in developed countries slumped 32 per cent to $132 billion (BNEF, 2013).

Figure 4.4 disaggregates global investments in renewables for different technologies between 2004 and 2012. It shows that investments mostly went to wind and solar, and that the proportion accounted for by solar investment overtook that for wind after 2009.

In Europe, cumulative investments resulted in a substantial increase of renewable electricity from 12.2 per cent in 1990 to 19.6 per cent in 2010,[4] with growth accelerating after 2004. Although hydro-power is still the largest renewable power category (55 per cent in 2010), growth has been particularly strong in wind, bio-power, and solar (which started from a small base). The share of renewable electricity varies widely between European countries,[5] with some high shares (in 2010) in countries with significant hydro-power (e.g. Norway (97.3 per cent), Austria (65.5 per cent), Latvia (42 per cent), Slovenia (32.2 per cent)) and wood/biomass (e.g. Sweden (56 per cent) and Finland (27.6 per cent)). In 2010, 'new' renewables made substantial contributions to power generation in Portugal (41.2 per cent), Denmark (32.9 per cent),

[4] See <www.eurostat.ec.europa.eu> (accessed May 2014).
[5] See <www.eurostat.ec.europa.eu> (accessed May 2014).

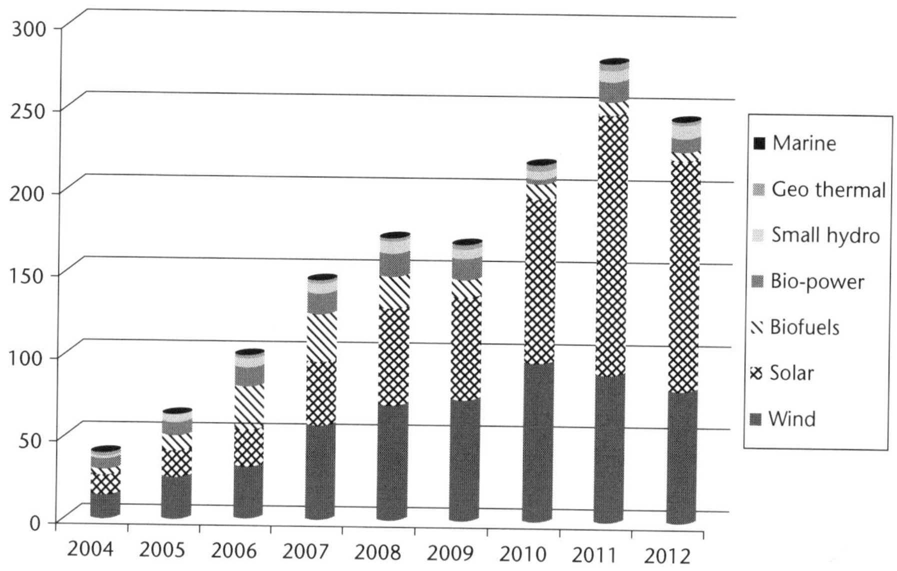

Figure 4.4. World renewable-based power sector investment by type (excluding large hydro) and total generation in $bn

Source: Constructed from data in BNEF, 2013

Spain (29.5 per cent), and Germany (18.1 per cent). Countries that lagged behind their targets were the Netherlands (9.7 per cent), the UK (7.4 per cent), Poland (6.7 per cent), and Cyprus (1.4 per cent).

An important driver for increasing interest in renewables is the decrease in cost. PV-modules, for instance, decreased from $4 per Watt in 2008 to less than $1 per Watt by January 2012 (Aanesen et al., 2012). Costs for onshore wind turbines fell by 15 per cent between 2009 and early 2013, while costs for offshore wind turbines rose 44 per cent as developers moved into deeper waters (BNEF, 2013). These price decreases partly reflect learning-by-doing and economies of scale, but mainly stem from increased production in China, which caused oversupply and price-dumping.

While the overall trend of cost decreases in renewable power is beyond doubt, precise numbers are uncertain and contested. One reason is that solar-PV systems require not just solar panels, but also inverters, interconnection wiring, sometimes a battery and solar track, and rooftop-integration materials, which have experienced limited price decreases (Zehner, 2012).[6]

[6] Zehner (2012) further suggests that solar cells may survive twenty to thirty years but that inverters generally last only five to eight years. Inverter replacement costs (and maintenance costs more generally) tend not to be included in cost estimates of solar power.

The same argument applies to wind turbines, which are only one component in wind parks. Another reason is that renewable electricity creates additional system costs and back-up costs: (a) because wind farms tend to be in remote locations, new transmission lines need to be built; (b) because wind and sunshine are intermittent, back-up capacity is needed (often gas turbines, which de facto leads to hybrid fossil fuel-renewable systems); and (c) addressing the intermittency issue also requires increased grid-monitoring and flexibility, which assumes investment in intelligence and control devices. These system costs increase rapidly when renewable capacity crosses a threshold of 20 per cent (Helm, 2012). Both kinds of additional costs complicate precise assessments of cost decreases in renewable electricity.

Despite cost decreases, renewable electricity options are still more expensive than power from coal or gas (and much more expensive for solar-PV and offshore wind, but approaching grid parity for onshore wind). The expansion of renewable electricity has therefore depended on regulations (e.g. renewable obligations), financial incentives (e.g. feed-in tariffs), and wider energy policy plans. An example of the latter is the Europe 20-20-20 policy with three targets for 2020: a 20 per cent share of renewables in *energy* (i.e. not just electricity) consumption, a 20 per cent reduction of GHG emissions, and a 20 per cent reduction in primary energy use through improved energy efficiency. Policy plans in other countries have also created favourable contexts for renewable energy, for instance, the UK Climate Change Act (2008), the UK Low-Carbon Transition Plan (2008), South Korea's Green Growth Strategy (2009–13), the German Energy Transition Plan (2010), India's National Solar Mission (2010), and China's Twelfth Five-Year Plan (2011–15).

These policy plans and instruments were developed in the context of rising public concerns about climate change. An analysis of British newspaper counts (Figure 4.5) shows a sharp rise in public attention to climate change from 2004 to 2007–8, fuelled by events such as Hurricane Katrina (2005), Al Gore's movie *An Inconvenient Truth* (2006), the Stern Review (2006), and the Fourth IPPC Assessment Report (2007). Public attention has subsequently decreased, not least because of the economic crisis and concerns about austerity and unemployment.

Another positive trend, and one that is often overlooked by national governments, is that many city and local governments are working on green reconfigurations of local energy, transport, and housing systems. REN-21 (2013: 77) documents many such initiatives and concludes that: 'Thousands of cities and towns around the world have active plans and policies to advance renewable energy. Despite the slowdown at the national level in 2012, policy momentum continued to accelerate at the local level as city governments took actions to generate employment, plan for rising energy demand, cut carbon emissions, and make cities more liveable.' In the US, more than 900 cities

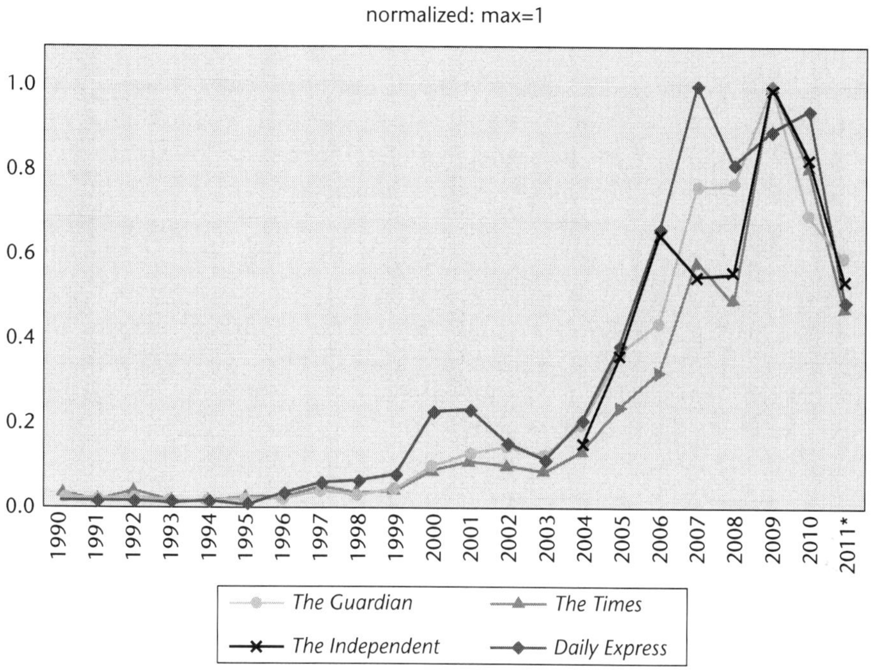

Figure 4.5. Yearly number of articles in UK national newspapers containing the phrase 'climate change'

had signed up (by 2009) to the Climate Protection Agreement (2005), launched by the US Council of Mayors, which requires signatories to meet US Kyoto targets (a 7 per cent reduction by 2012 on 1990 levels) within their own boundaries. In 2005, global megacities (e.g. Berlin, Hong Kong, Jakarta, Johannesburg, Los Angeles, London, New York City, Tokyo) created the C40 Cities Climate Leadership Group (2005), which aims to reduce GHG emissions and adapt to climate change. In Europe, there are various dedicated urban initiatives (e.g. Copenhagen's Plan to be carbon-neutral by 2025), including the development of eco-cities (e.g. Freiburg, Graz). These urban initiatives are encouraging, first because they entail on-the-ground implementation and reconfiguration of concrete systems. Second, city governments may prove to be a new kind of actor that is less constrained by existing regimes (see later in the chapter), and has the capabilities and resources to drive transitions on-the-ground.

Negative Developments in Low-Carbon Electricity Transitions

While the expansion of renewable electricity niches is substantial, there are also various developments that complicate or run counter to low-carbon transitions. The momentum of niche-innovations appears to have slowed down, as the financial-economic crisis has weakened some of the drivers behind renewables (Geels, 2013). First, the crisis has decreased public concerns about climate change (Figure 4.5). A public attitudes survey (March 2012) by the UK Department of Energy and Climate Change (DECC) found that people worry more about economic issues than about energy and climate change. Similar shifts in public concerns can be found in other countries. Second, the crisis has weakened green policies, as austerity programmes led many European governments (Germany, UK, Spain, Italy) to reduce feed-in tariffs for renewable electricity (BNEF, 2013). The drive to reduce public spending also led many countries to abandon green stimulus packages, which were introduced in the early crisis years to stimulate 'green growth'. Since 2012, this has resulted in a decline in public grants, cheap loans, and investments in green technologies and sectors. Other policy pressures also weakened. The international climate negotiations have stalled, with countries agreeing in Durban (in November 2011) to delay further talks until 2015, when they will discuss a possible international treaty that could come into force in 2020. Furthermore, the European Emissions Trading System (ETS), the EU's flagship policy instrument, is not driving low-carbon transitions, because the carbon price is relatively low, decreasing, and volatile. Third, the crisis has, since 2012, led to declining global investments in renewable energy options, particularly in Europe and the United States (see Table 4.2 and Figure 4.4), because of austerity pressures and investor uncertainty, related to reduced government support. In summary, the financial-economic crisis has weakened some of the main driving factors, which may result in setbacks and slow down the low-carbon transition. The discourse of 'green growth' created a window of opportunity in the early crisis years (2008–10), resulting in a number of green stimulus packages and investments in renewable electricity. Since then, however, this window has been shrinking (Geels, 2013) as austerity programmes have cut public spending and socio-political attention has shifted towards the costs of low-carbon transitions.

Low-carbon transitions are also hindered or complicated by ongoing developments in existing regimes. The remainder of this section will sketch some of the main trends in natural gas, nuclear, and coal regimes at the global level, and section 4.4 will provide a more focused analysis of the UK.

An important development in the gas regime is the 'shale gas revolution' in the United States over the last five years, based on fracking gas trapped in underground rocks. This 'revolution' has caused a steep decline in US gas prices (Figure 4.6), which, in turn, has stimulated utilities to (partly) switch from coal to gas. Other countries (e.g. China, UK, Canada, Australia) are also exploring the potential for shale gas, which could be a 'game changer' since estimates of world gas reserves have doubled in just a few years. Reserves in the US are now estimated to be sufficient to produce gas for 100 years (Helm, 2012).

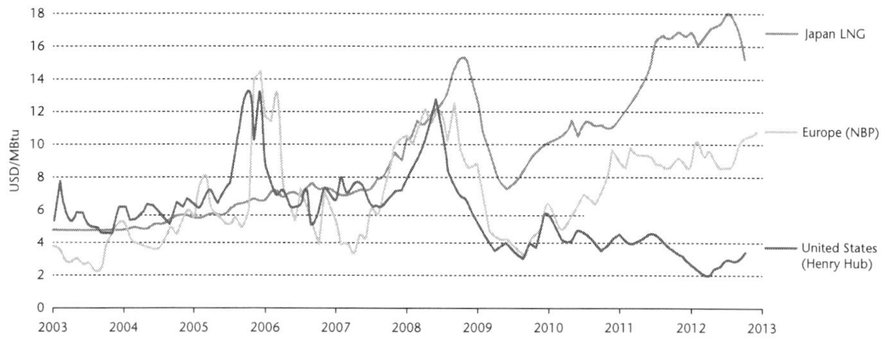

Figure 4.6. Natural gas spot prices

Source: IEA, 2013b, 42; (NBP = National Balancing Point (UK) representative of European gas prices)

The IEA (2011) predicts a 'golden age for natural gas', suggesting that global gas demand may rise by more than 50 per cent between 2010 and 2035, outstripping the growth in renewables. This suggests that gas may be a double-edged sword. On the one hand, a coal-to-gas shift may substantially reduce CO_2 emissions, as in the US. On the other hand, increasing gas use may hinder the progress of renewable energy and possibly kill off the recent investment wave, which is likely to damage investor confidence for many years to come.[7] Additionally, natural gas is still a fossil fuel, although less polluting than coal. And shale gas exploitation may have negative implications (e.g. ground water pollution by chemicals used for fracking, earthquake tremors, disruption of local villages). An indirect effect is that cheap US coal has flooded global markets, stimulating increased coal use elsewhere.

Developments in the nuclear power regime are ambiguous in at least two ways. The first ambiguity is whether or not nuclear power is desirable. On the one hand, it is a low-carbon option, which is partly why many governments and advisory bodies (IPCC, IEA) see building new plants as a component of low-carbon scenarios. The IEA's (2013a) 2° scenario, for instance, assumes that nuclear power generation needs to increase from 2518 TWh in 2011 to about 4500 TWh in 2025. This doubling of capacity is quite a challenge, because nuclear generation has stagnated since the 1990s, and decreased from 2600 TWh in 2010 to 2346 TWh in 2012 (Schneider and Froggatt, 2013). On the other hand, many people and organizations dislike nuclear energy, because of

[7] The IEA (2013b: 40) suggests that 'access to investment capital for renewables may be tighter due to competition from gas'.

nuclear waste problems, the risk of accidents and nuclear weapons proliferation, and dubious economic viability; because of these reasons, many privatized electricity companies have not invested in nuclear power since the 1990s.

The second ambiguity is that there are conflicting decommissioning and construction trends. On the one hand, various countries (Germany, Belgium, Switzerland, collectively accounting for about 30GW capacity) decided to phase out nuclear power, after the 2011 Fukushima nuclear accident. Japan is currently considering its dependency on nuclear power, and the French Government is considering reducing the share of nuclear electricity from 79 per cent to 50 per cent by 2025 and increasing the share of renewables (IEA, 2013b). On the other hand, countries like the United States, China, India, Russia, and the UK have active or planned nuclear expansion plans. In 2012, there were sixty-eight reactors under construction globally, two-thirds (44) of which are located in just three countries (India, Russia, China). These new reactors would boost capacity by 67 GW.[8] New capacity would take a long time to come on line, because average construction times are five to ten years (and delays are frequent), not including planning and consulting. Even if all the plans were realized, the IEA (2013b) warns that nuclear expansion is proceeding too slowly compared to their 2° scenario.[9] It therefore suggests that major new construction programmes are needed, but also realizes that 'this will require greater public acceptance of nuclear energy, and more favourable electricity market mechanisms and investment conditions' (2013b, 32).

Another important development is the global expansion of coal-fired power generation, which increased by 45 per cent between 2000 and 2010 and by 6 per cent between 2010 and 2012, faster than non-fossil energy sources on an absolute basis. 'For all the talk about natural gas and renewables, coal unquestionably won the energy race in the first decade of the 21st century' (IEA, 2011, 355). Coal is globally by far the largest source of electricity generation, increasing its share from 39 per cent in 2000 to 42 per cent in 2010 (IEA, 2013b). Countries with high coal shares in power generation (in 2010) are South Africa (93 per cent), Poland (90 per cent), China (79 per cent), Australia (70 per cent), India (69 per cent), the US (45 per cent), and Germany (44 per cent) (Helm, 2012). China's coal consumption represented 46 per cent of global demand in 2011, while India's share was 11 per cent. Cheap coal from the US and low carbon prices in the ETS also stimulated European coal use in recent years. In 2012, coal use in power generation expanded by 2.8 per cent in

[8] Schneider and Froggatt (2013) place some caveats about too much optimism, remarking that thirteen reactors have been labelled 'under construction' for more than ten years, and that forty-five projects do not yet have an official start date.

[9] If all the announced plans were realized, nuclear capacity would increase from 392 GW in 2012 to between 440 and 555 GW in 2025, which would still fall short of the IEA's (2013b) 2° scenario in which nuclear capacity should be about 650 GW in 2025.

Europe, 6 per cent in Germany, and 32 per cent (!) in the UK (IEA, 2013a, and Figure 4.7). This increase is an indirect effect of the US shale gas revolution, which led to increasing US coal exports and decreasing coal prices, from $130/ton in March 2011 to $85/ton in May 2012 (IEA, 2013b).

Coal-fired power expansion is a major reason why the Energy Sector Carbon Intensity Index (ESCII), i.e. the amount of CO_2 emitted for each of unit of energy supplied, has remained essentially static, changing by less than 1 per cent since 1990 (IEA, 2013b).[10] Hence, despite the growth of renewables and natural gas, power generation is almost as carbon-intensive as twenty-three years ago. Maintaining current policies would see coal use rise by a further 65 per cent by 2035 (IEA, 2011, 5), which would put the 2° target out of reach.

The continued coal expansion 'presents a major threat to a low-carbon future' (IEA, 2013b). The development and implementation of carbon capture and storage (CCS) could perhaps provide a solution to this problem because it would enable coal (and gas) burning without CO_2 emissions. Unfortunately, the IEA (2013a, 13) assesses that 'large-scale carbon capture and storage remains distant'. Current CCS capacity captures only 6 million tonnes (Mt) of CO_2 per year. There are provisions to contract CCS-capacity for a further 13 Mt CO_2 in 2013, and more distant plans which would bring the total to 90 Mt CO_2. But this would still capture less than 1 per cent of power-sector CO_2 emissions in 2012 (IEA, 2013a). The IEA director therefore expresses serious concerns: 'I am particularly worried about the lack of progress in developing policies to drive carbon capture and storage (CCS) deployment. Without CCS, the world will have to abandon its reliance on fossil fuels much sooner, and that will come at a cost' (IEA, 2013b, 5). Alternatively, and perhaps more likely at present, the world will continue to burn fossil fuels and transgress the 2° target.

4.4. The Political Resilience of UK Gas, Nuclear, and Coal Regimes

Section 4.3 highlighted that the expansion of existing regimes forms a major obstacle to low-carbon transitions. This section will show that this problem not only applies to developing and emerging economies (e.g. China, India, South Africa), but also to developed countries such as the UK. In particular, this section will further explore some of the political dimensions to the resilience and resistance of existing regimes. The focus is on one country, the UK, which enables us to carry out a more detailed analysis of

[10] In contrast, ESCII improved by 6% between 1971 and 1990, following the 1973 and 1979 oil shocks (IEA, 2013b).

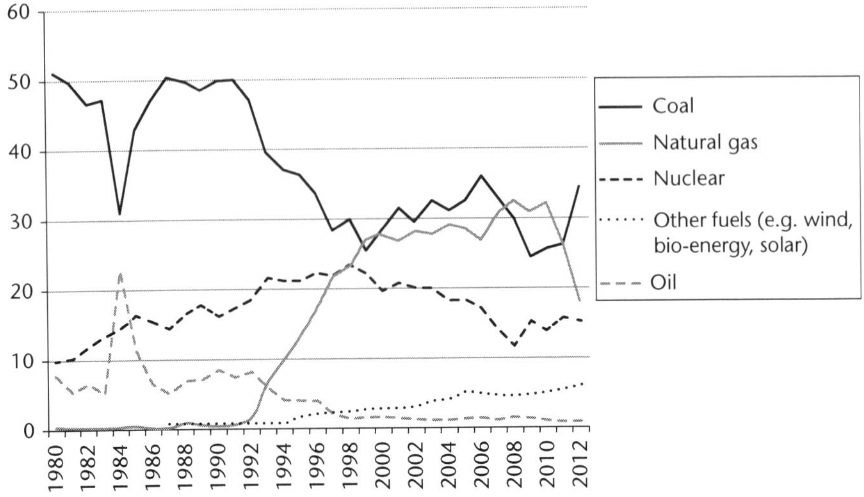

Figure 4.7. Fuel input to central power generation in the UK, million tons of oil equivalent

Source: Constructed from data from DUKES, Digest of UK Energy Statistics, https://www.gov.uk/government/collections/digest-of-uk-energy-statistics-dukes

two incumbent actors in the upstream electricity production regime: electric utilities and governments.[11] Three subsequent sections will analyse how their commitments and strategies shaped the development of three technologies in the electricity regime: gas, nuclear, and coal.[12] Figure 4.7 provides a longitudinal overview of their relative importance in power generation over the last thirtry-two years.

Natural Gas

Until 1990, natural gas was a premium fuel, used for domestic heating and cooking and for peak demand in electricity production. During the 1990s, the UK experienced an unanticipated 'dash for gas', with gas being used for

[11] Other important actors, which receive less attention in this chapter, are grid operators, electricity distribution companies, large industrial users (who often generate their own power), households, fuel suppliers, and equipment manufacturers (e.g. Siemens, General Electric).

[12] The analysis draws on Geels (2014). The text in this chapter is organized more systematically around three technologies, whereas the Geels (2014) article presents a running conceptual argument (about different kinds of regime resistance), illustrated with ad hoc examples from the UK electricity domain.

base-load generation in combined cycle gas turbines (CCGT). The dash for gas was stimulated by the confluence of various factors (Winskel, 2002): (a) privatization (1990) freed the electricity supply industry from regulated use of UK coal, enabling it to choose its own fuel inputs; (b) changing investment conditions and a preference of utilities for smaller-size units with short lead-times, low capital costs, and quick returns on investment—criteria that matched CCGT which was essentially modular and could be built faster (in about one year) than large coal-fired plants; (c) price/performance improvements in CCGT; (d) gas had environmental benefits over coal, which made it easier to meet European environmental regulations (regarding sulphur emission and acid rain); and (e) the availability of cheap natural gas from new North Sea gas finds and international markets. The rise of gas caused such a rapid decline in coal-fired generation that the UK Government issued a moratorium (in 1997) to maintain fuel diversity (Winskel 2002). The decline in North Sea gas production after 2000 has been compensated by increasing gas imports, which in 2012 accounted for half of gas consumption.

The 1997 Kyoto Protocol increased the political salience of climate change and the Labour Party made the issue a central component of its 1997 election campaign. An influential report by the UK Royal Commission on Environmental Pollution in 2000 further catalysed the climate change debate, suggesting the need for a 60 per cent reduction of GHG emissions by 2050. The 2003 White Paper, *Our Energy Future: Creating a Low-Carbon Economy*, made climate change a central policy issue and embraced the importance of renewable energy.

In subsequent years, however, policy thinking and priorities changed because oil and gas prices rose gradually after 2000 and more rapidly after 2004 (Figure 4.8). Coupled with Russian gas supply problems in 2005, rising prices created new concerns for utilities and policy-makers about energy security and affordability. In 2005, the British Prime Minister, Tony Blair, announced a review of UK energy policy, arguing that 'Round the world, you can see feverish rethinking. Energy prices have risen. Energy supply is under threat. Climate change is producing a sense of urgency' (reported in *The Independent*, 30 November 2005). The 2007 White Paper (*Meeting the Energy Challenge*) established prices and energy security as strategic policy issues, along with climate change: 'We need therefore to establish a strategy which delivers both energy and climate security. It is not sustainable to achieve one without the other' (p. 28). The White Paper also broadened government support to three technological options: renewables, CCS for fossil fuels (coal, gas), and nuclear power. Coal and nuclear power thus returned to the policy agenda.

Faced with market uncertainties and rising gas prices, utilities diversified and from 1999 to 2006 increased their coal use (Figure 4.7). In the last

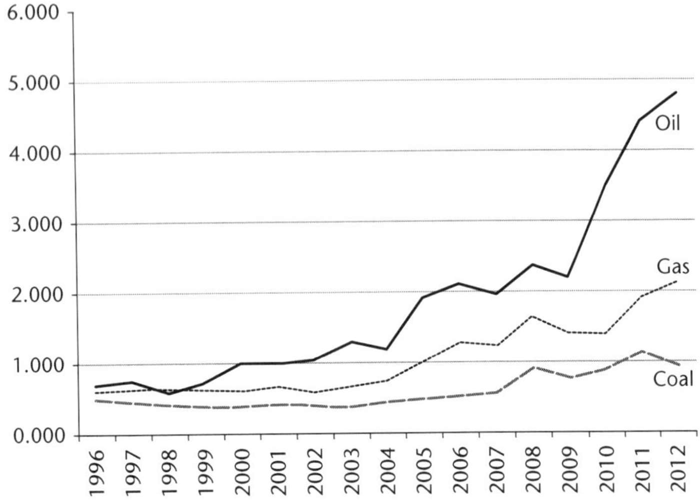

Figure 4.8. Average fuel prices (pence per kWh) paid by UK power producers
Source: Constructed from data from DUKES, Digest of UK Energy Statistics, https://www.gov.uk/government/collections/digest-of-uk-energy-statistics-dukes

few years, the US shale gas revolution and associated price decreases triggered interest in the UK. In December 2012, the UK Government lifted restrictions on fracking and dozens of potential shale gas sites have since been identified across the country. In July 2013, the Chancellor outlined plans to give generous tax breaks to shale gas companies, arguing that he wants Britain to become a leader of the shale gas revolution 'because it has the potential to create thousands of jobs and keep energy bills low for millions of people'. In 2013, the Government also signalled its ambition to see the building of forty new gas-fired power stations, which gave rise to a new controversy. On the one hand, domestic shale gas would contribute to energy security and could diminish GHG emissions from coal. On the other hand, the UK Committee on Climate Change has warned that a big expansion in gas use would be incompatible with the carbon targets formulated in the 2008 Climate Change Act. Conservative politicians are therefore attempting to weaken these targets, which were enshrined in law. Already at the Conservative Party conference in October 2011, for instance, Chancellor George Osborne argued that the UK should not pursue green policies that would 'pile costs on the energy bills of households and companies' and might put the country 'out of business'. He further suggested that UK's post-2020 carbon targets should be subject to review, and pledged that the UK should not cut carbon emissions faster than other European countries. In this context, it is not insignificant that the new Energy Bill (November 2012) did not set a long-term decarbonization target for GHG emissions from power stations. Shale gas became the topic of a heated controversy in the second half of 2013, when the Government dismissed local opposition and protests against fracking, and expressed its firm commitment to shale gas (e.g. in a letter to the *Daily Telegraph* on 11 August 2013, written by the Prime Minister). The Government aimed to overcome local protests by suggesting that local authorities should receive 1 per cent of shale gas revenues derived in their area.

Nuclear Power

The UK Government has stimulated and subsidized nuclear power since the 1950s. In the late 1980s, preparation for electricity sector privatization revealed the poor economic performance of nuclear power, which is why nuclear power plants were withdrawn from the sale. In 1989, the Government also announced a moratorium on new nuclear power plants. Scandals with regard to the storage of nuclear waste (Sellafield) and reprocessing (THORP) further undermined the cultural legitimacy of nuclear power (Verhees, 2012).[13] The 2003 White Paper (*Our Energy Future*) seemed to seal nuclear energy's fate, proclaiming that: 'Its current economics make nuclear power an unattractive option for new, carbon-free generating capacity and there are also important issue of nuclear waste to be resolved.' The White Paper further promised that decisions to build new nuclear power stations would require 'the fullest public consultation'.

In 2005, however, nuclear power reappeared on the agenda as a possible answer to climate change and energy security. Sir David King, the Government Chief Scientific Adviser, argued that: 'We need indigenous sources so we don't rely on imported gas from Russia... Indigenous supplies include all renewables and nuclear. Relying on renewable sources to replace lost capacity when existing nuclear power stations close would be a tough challenge' (reported in *The Guardian*, 21 October 2005). The Prime Minister Tony Blair also began to advocate nuclear power, and in 2006 the Department for Trade and Industry published an energy review (*The Energy Challenge*) which announced plans to build new nuclear stations over the next two decades. In 2006, the Government's Environmental Audit Committee warned that new nuclear plants would not be able to solve the energy gap (which would result from the decommissioning of old coal and nuclear plants by 2015/2016), because of long lead-times. In 2007, Greenpeace won a legal challenge against the energy review, on the basis that the Government had failed to engage in

[13] These scandals included unplanned waste discharges, falsified quality control documents, and suspect shipments to Germany and Japan (Verhees, 2012).

'the fullest public consultation'. Although the Government subsequently launched a new consultation, Blair announced in advance that he would not listen and that 'This won't affect policy at all'. Indeed, the 2007 White Paper (*Meeting the Energy Challenge*) and the 2008 report (*Meeting the Energy Challenge: A White Paper on Nuclear Power*) positioned nuclear power as a necessary option to address energy security and climate change (besides renewables and CCS for fossil fuels). In 2008, the new Prime Minister, Gordon Brown, called for a 'nuclear renaissance' and the construction of at least eight nuclear power plants (Verhees, 2012). In 2009, the Government published a list of eleven potential sites for new nuclear plants, which was reduced to eight in 2010.

The debate about nuclear power caused a schism in the environmental movement, with some activists (e.g. former Greenpeace director, Stephen Tindale) perceiving nuclear power as a necessary evil to address climate change. This schism weakened public opposition to nuclear power, which helps explain why the 2011 Fukushima nuclear accident did not derail the Government's plans. A larger problem was the lukewarm interest from private companies (utilities), which would actually have to build and operate the nuclear plants. The unfavourable economics of nuclear power, waste-processing liabilities, and decommissioning costs created major uncertainties about the viability of nuclear investments, especially since the Government had repeatedly stated that it would not provide public subsidies for nuclear power. In February 2013, the energy company Centrica abandoned plans to build new nuclear power plants, leaving only EDF (Electricité de France) in concrete negotiations with the UK Government about a new plant at Hinkley Point. To sweeten the deal, the Government agreed in late 2013 to pay EDF a high guaranteed electricity price (twice the current wholesale price) for thirty-five years. Although this deal was called a 'contract for difference', critics argue that this label is a ploy to hide the fact that the Government has broken its pledge not to subsidize nuclear reactors.

Coal

The dash for gas caused rapid declines in coal use for power generation. Although the government moratorium (in 1997) on gas-fired power plants created some protection, there were no plans to build new coal-fired plants. Coal also faced environmental pressures, initially from acid rain and later from climate change. When the acid rain issue emerged in the 1970s, the British Government and the electricity industry highlighted scientific uncertainties and costs to justify inaction. Throughout the 1980s, the Thatcher Government frustrated and hindered international discussions, which, together with high sulphur emissions, gained Britain the reputation of the 'dirty man of

Europe' (Turnheim, 2012). The British position changed in the late 1980s. In 1988, Thatcher gave a speech that acknowledged the problems of climate change and acid rain. This was followed by agreement on the European Large Combustion Plants Directive (LCPD) prescribing progressive reductions in SO_2 emissions, which was assumed to imply the installation of flue gas desulphurization devices. The UK, however, did not immediately translate the LCPD into law or technological mandates, but instead took advantage of the dash for gas to achieve the targets at low cost. In the 1990s, increasing concerns about climate change created additional pressure on coal. As a result of these various factors, coal was perceived to be on its way out. Indeed, 'nobody saw any coal revival in the EU, only further decline' (Smil, 2003, 236).

By the early 2000s, however, coal reappeared on the agenda for three reasons. First, rising oil and gas prices caused utilities to use more coal between 1999 and 2006 (Figure 4.7). In 2007–10, decreasing gas prices (Figure 4.8) stimulated gas use and decreased coal use. Since 2010, however, coal has bounced back (Figure 4.7) because of increasing gas prices (Figure 4.8) and decreasing coal prices.

Second, concerns about energy security increased the interest in coal, leading to political support for clean-coal initiatives. The Trade and Industry Secretary explained that the UK would not place all its eggs in one basket: 'If a new, cleaner coal generation is viable, then I think it could have an important part to play in making sure we have diverse generation in the future. Coal is easy to store and it comes from a variety of well-established sources around the world' (reported in *The Guardian*, 21 February 2006). Interest in coal also increased because policy-makers realized that the approaching energy gap (related to the decommissioning of old coal-fired plants in 2015 because of the LCPD) would be difficult to fill with renewables and nuclear power (because of long lead-times). The 2007 White Paper (*Meeting the Energy Challenge*) therefore acknowledged (clean) coal as part of the solution to climate change and energy security. Utilities also embraced coal: 'Interest in new coal plants has been sparked by relatively low coal prices over the last five years, making them more profitable than gas plants. However, the strongest driver for new coal is the desire by companies to maintain a portfolio of generating capacity that includes a range of fuels, to hedge against markets, security of supply or policy risk' (Lockwood, 2008, 6). In 2006, the energy firms E.On and RWE announced intentions to build the first UK coal plants in twenty years to compensate for the closure of capacity under the LCPD (Turnheim, 2012). These plans were little influenced by the European Emissions Trading Scheme. By 2008, various coal-fired projects totalling just over 11GW were being planned or pending approval. However, complicated planning permissions and environmental activism have delayed these projects (Turnheim, 2012).

The third contributor to coal's revival was the discourse of 'clean coal', which related to technical innovations such as flue-gas desulphurization devices, supercritical pulverized coal technologies, and coal gasification. The most important innovation, however, was carbon capture and sequestration (CCS), which promised to diminish GHG emissions substantially. The 2007 White Paper announced a CCS demonstration programme supported by a £1 billion subsidy (Turnheim, 2012). The promise of CCS also provided legitimation of the White Paper's endorsement of future coal plants. But although CCS is technically feasible, there is much uncertainty about its commercial viability. Utilities are not keen to deploy the technology and look for government support and hand outs. Most of the corporate activity on CCS has therefore 'been focused on basic scientific research and lobbying governments for subsidies and support rather than investments needed to deploy the technology on a commercial scale' (Bowen 2011, 2256). In 2008, four consortia (Peel Energy, BP, E.ON, and Scottish Power) showed interest in the subsidized demonstration programme. By 2010, however, firms began pulling out. The last consortium (Scottish Power) disbanded in October 2011, creating serious problems for the Government's energy strategy (which assumes widespread future use of CCS).

Opponents have voiced concerns that CCS is a technological promise that will never materialize and is being used as an argument to delay carbon abatement. The latest part of this controversy is the proposal by CCS proponents that new-built coal plants can be designed to be 'capture-ready', which seems to imply the intention of firms to add CCS to coal-fired plants when it becomes feasible in the future (Turnheim, 2012). But opponents see this as a flimsy promise, which the industry is using to get permits to build new coal plants. They fear that 'capture-ready' plants may never retrofit CCS because of the high costs involved. An additional argument is that investment in clean coal (and nuclear) will diminish the future space for renewables: 'The government's system [the policy framework for energy, FWG] is designed to accommodate large, inefficient and remote power stations owned by large companies like the coal and nuclear utilities such as E.ON. In other words, nuclear and coal power stand like two bouncers at the door blocking the way for renewables and efficiency, and perpetuating our outdated, inefficient and centralised energy system' (Greenpeace 2008, 9–10).

4.5. Conclusions

The global low-carbon transition in electricity has made some progress, but also faces major obstacles. Using the MLP to make a general assessment of overall patterns, the following can be concluded. On the one hand, the first part of section 4.3 showed that several renewable electricity niche-innovations (wind, solar-PV, biomass) have gained some momentum (especially in Europe and China), and that there are encouraging city initiatives and developing country investments. On the other hand, the second part of section 4.3 showed that existing regimes are relatively stable (in terms of incumbent actor commitments) and economically expanding (in the case of gas and coal). This relative regime stability means that 'windows of opportunity', which the MLP suggests are necessary for the wider breakthrough of green niche-innovations, are limited. So, while renewable energy options may grow in certain niche applications, they face an uphill struggle against existing regimes (especially coal and gas) in mainstream markets. The nuclear power regime is more ambiguous in the sense that policy-makers in some countries (e.g. Germany, Switzerland) have defected from it. This nuclear regime destabilization creates windows of opportunity for green niche-innovations, but also threats. In Germany, for example, the *EnergieWende* envisages a transition to renewable energy. But the decommissioning of nuclear power also creates opportunities for coal-fired power, which actually grew by 6 per cent in 2012.

Section 4.4 showed in more detail that incumbent regime actors will not roll over because of climate change concerns. Instead, they defend themselves by adapting their innovation efforts and through political framing strategies. Section 4.4 showed that the 2003 UK White Paper (*Our Energy Future*) was a threat because it portrayed renewable electricity as the main low-carbon option. But regime actors successfully adapted: nuclear actors framed their product as 'low-carbon' (sidelining other environmental considerations) to make climate change into an opportunity for future expansion, while coal and gas actors used the promise of CCS to convince policy-makers and other elites of their future viability in a low-carbon world. Incumbent regime actors such as the UK Government and utilities appear to be firmly committed to gas and clean coal (with the promise of CCS). The government would also like to see nuclear expansion, but investors and utilities are still hesitant because of various uncertainties. The analysis in section 4.4 also shows that climate change is not the only consideration of the government, utilities, and investors. Other factors such as energy security, affordability, diversity, flexibility, fuel costs, and return on investment are also (and probably more) important for these actors, which does not bode well for low-carbon transitions. Additionally, UK public concerns about climate change have weakened since 2010 (Figure 4.5), and the costs of renewables have become a more prominent issue in public discourse. Government support for renewables is not getting stronger, as indicated by the reduction of feed-in tariffs and the lack of a long-term decarbonization target in the 2012 Energy Bill. Because of these considerations, renewable electricity will face an uphill struggle in the UK against resilient regimes. It is far from certain that the UK will meet the carbon targets from the 2008 Climate Change Act, or that this Act itself will survive unamended (Lockwood, 2013).

On a global scale, it also looks unlikely, on present trends, that the world will meet the 2° target.[14] The continued expansion of fossil-fuel regimes (particularly coal) is a major problem, because it rapidly consumes the remaining 'carbon budget'. Climate scientists have calculated that CO_2 concentrations need to stay below 450 parts per million (ppm) to have a 50 per cent chance of remaining within the 2° target (Berners-Lee and Clark, 2013). The remaining 'carbon budget' associated with 450 ppm is about 1440 Gt CO_2, of which 420 Gt has already been emitted between 2000 and 2011 (IEA, 2013a). Because another 136 Gt CO_2 is estimated to be emitted from non-energy related sources before 2050 (e.g. from agriculture), the remaining budget is about 884 Gt CO_2 by 2050 (IEA, 2013a). The central problem is that the world's *proven* fossil-fuel reserves are much larger than the 884 Gt carbon budget, namely about 2800 Gt (IEA, 2013a), with about 1870 Gt coal, 610 Gt oil, and 410 Gt gas.[15] This means that the low-carbon transition challenge is not just to stimulate 'green' alternatives, but also to actively manage the decline of fossil-fuel-based systems and to prevent existing fossil-fuel reserves from being burnt (Berners-Lee and Clark, 2013). This will entail political struggles with fossil-fuel companies and other incumbent firms (e.g. electric utilities, car companies), who have invested in finding the fossil-fuel reserves and now plan to exploit those. The ongoing global coal expansion (recently also witnessed in various European countries) suggests, however, that there is limited political will to enter into these struggles. Instead, policy-makers place high hopes on 'green' innovation rather than also considering the management of the decline of existing systems. One problem with this orientation is that renewable electricity sources have so far been mainly *additional* to fossil fuels with no (or limited) substitution effects.[16] Another problem is that 'emergent' transitions (arising from uncoordinated activities and market-driven substitution) will probably unfold too slowly compared to the time scale suggested by climate scientists. These scientists suggest that global CO_2 emissions should start decreasing by 2020, because of lag-effects in the climate system and the cumulative nature of CO_2 emissions (Anderson and Bows, 2011). Compared to historical transitions, which often take several decades (Geels, 2005), this desired speed is a great challenge. More 'purposive'

[14] In other sectors (e.g. buildings, agro-food, transportation), there is generally less progress than in electricity.

[15] *Probable* reserves and *possible* reserves of fossil fuels are far larger still, maybe even twenty times larger than proven reserves for coal (Berners-Lee and Clark, 2013).

[16] This is not an uncommon pattern in transitions, however. Geels (2005) demonstrated that new technologies may initially piggyback on expanding markets, developing in tandem with existing technologies. Only after a few decades do niche-innovations enter mainstream markets and substitute for existing technologies.

transitions may therefore be required (Smith et al., 2005), with a more active role for governments.

The IEA (2013a, 9) warns that: 'the 2-degree target still remains technically feasible, though extremely challenging. To keep a realistic chance of meeting the 2-degree target, intensive action is required before 2020.' I would add, first, that current policies remain too wedded to the needs of incumbent actors, and that much bolder policies are required. Second, this chapter has highlighted that intensive policy action should not just consist of pushing green alternatives, but also of managing the decline of fossil-fuel-based systems (or stimulating the widespread adoption of CCS). This combined challenge, and not just support for innovation, is likely to become an increasingly important topic, both in politics and in academic debates on transitions.

The broader implication of both points is that discussions of climate change policy should not just focus on policy instruments, but also on problem definition, agenda setting, political economy, power struggles, and political will, which is what section 4.4 attempted to do. In this respect, I agree with Sovacool (2009, 1539), who distinguishes thirty policy mechanisms to stimulate renewable electricity and energy efficiency, discusses four in depth, and then concludes: 'What seems to be lacking is not the availability of robust public policy mechanisms, but the political and social will to implement them.'[17] The German Advisory Council on Global Change likewise concludes: 'The technological potential for comprehensive decarbonization is available . . . and the policy instruments needed for a climate-friendly transformation are widely known. Now it is foremost a political task to overcome the barriers of such a transformation, and to accelerate the change' (WBGU, 2011, 1).

With regard to political will, however, I suggest that a strategic political push-back may be under way, for instance in the UK, where politicians have opportunistically politicized rising energy costs, blaming either the pricing policies of energy companies (the Labour Party) or green policies (the Conservatives) (Geels, 2014). Compared to other European countries, however, UK electricity prices are about 19 per cent below the EU average (Bolton, 2013). Consequently, although prices have been rising rapidly in recent years (by about 8–10 per cent a year), they are still not very high in relative terms. This suggests that UK policy-makers deliberately chose to emphasize high energy prices, because it is an easy story that resonates with citizens (especially in

[17] These four policies are: to eliminate subsidies for mature energy technologies; to create accurate electricity prices (that do not obfuscate but are understandable, and internalize external costs); to create national feed-in tariffs (which create investor certainty) and guarantee renewable power generators access to the grid; and to inform and engage the public through broad public debates that convey positive visions. Other relevant policies are: to increase funding for energy R&D; to offer low-interest loans and/or government financing; to streamline permitting and siting procedures; and to pass renewable electricity obligations.

times of austerity), enabling them to score cheap political points. The immediate effect, however, has been that the UK Coalition Government has used the price controversy to downscale various green policies, which is likely to hinder the low-carbon energy transition. Political push-back is also visible in Japan, which reduced its commitment to GHG emission reductions, and in Australia, where the new Prime Minister (Tony Abbott) has scrapped the carbon tax and weakened other climate change policies. These worrying signs suggest that the political struggles related to climate change are far from over, and that counter-forces may substantially delay the progress towards low-carbon transitions, which climate scientists say is necessary to achieve the 2° target.

References

Aanesen, K., Heck, S., and Pinner, D. (2012). *Solar Power: Darkest Before Dawn*, New York: McKinsey on Sustainability and Resource Productivity report.

Anderson, K. and Bows, A. (2011). 'Beyond "Dangerous" Climate Change: Emission Scenarios for a New World', *Philosophical Transactions of the Royal Society A*, 369: 20–44.

Berners-Lee, M., and Clark, D. (2013). *The Burning Question: We Can't Burn Half the World's Oil, Coal and Gas. So How Do We Quit?*, London: Profile Books.

BNEF (2013). *Global Trends in Renewable Energy Investment 2013: Key Findings*, Frankfurt am Main: Bloomberg New Energy Finance, in collaboration with Frankfurt School of Finance and Management.

BNEF (2014). *Global Trends in Renewable Energy Investment 2014*, Frankfurt am Main: Bloomberg New Energy Finance, in collaboration with Frankfurt School of Finance and Management.

Bolton, P. (2013). *Energy Prices*. Research briefing for the House of Commons Library, <www.parliament.uk/business/publications/research/briefing-papers/SN04153/energy-prices>, accessed Mar. 2014.

Bowen, F. (2011). 'Carbon Capture and Storage as a Corporate Technology Strategy Challenge', *Energy Policy*, 39: 2256–64.

Elzen, B., Geels, F. W., and Green, K. (eds) (2004). *System Innovation and the Transition to Sustainability: Theory, Evidence and Policy*, Cheltenham: Edward Elgar.

Freeman, C., and Perez, C. (1988). 'Structural Crisis of Adjustment, Business Cycles and Investment Behaviour', in G. Dosi, C. Freeman, R. Nelson, G. Silverberg, and L. Soete (eds), *Technical Change and Economic Theory*, London: Pinter, 38–66.

Geels, F. W. (2002). 'Technological Transitions as Evolutionary Reconfiguration Processes: A Multi-Level Perspective and a Case-Study', *Research Policy*, 31(8–9): 1257–74.

Geels, F. W. (2005). *Technological Transitions and System Innovations: A Co-evolutionary and Socio-Technical Analysis*, Cheltenham: Edward Elgar.

Geels, F. W. (2012). 'A Socio-Technical Analysis of Low-Carbon Transitions: Introducing the Multi-Level Perspective into Transport Studies', *Journal of Transport Geography*, 24: 471–82.

Geels, F. W. (2013). 'The Impact of the Financial-Economic Crisis on Sustainability Transitions: Financial Investment, Governance and Public Discourse', *Environmental Innovation and Societal Transitions*, 6: 67–95.

Geels, F. W. (2014). 'Regime Resistance Against Low-Carbon Energy Transitions: Introducing Politics and Power in the Multi-Level Perspective', *Theory, Culture and Society*, 31(5): 21–40.

Geels, F. W., and Schot, J. W. (2007). 'Typology of Sociotechnical Transition Pathways', *Research Policy*, 36(3): 399–417.

Greenpeace (2008). *The Greenpeace Case Against Coal-Fired Power Generation*, Greenpeace UK, <www.greenpeace.org.uk/media/reports/the-case-against-coal-fired-power-generation>, accessed Mar. 2014.

Helm, D. (2012). *The Carbon Crunch: How We're Getting Climate Change Wrong and How to Fix It*, New Haven and London: Yale University Press.

International Energy Agency (2011). *World Energy Outlook*, Paris: IEA.

International Energy Agency (2013a). *Redrawing the Energy Climate Map, World Energy Outlook Special Report*, Paris: IEA.

International Energy Agency (2013b). *Tracking Clean Energy Progress 2012: IEA Input to the Clean Energy Ministerial*, Paris: IEA.

IPCC (Intergovernmental Panel on Climate Change) (2007). *Fourth Assessment Report*, Cambridge: Cambridge University Press.

Lockwood, M. (2008). *After the Coal Rush: Assessing the Policy Options for Coal-Fired Power Generation*, London: Institute for Public Policy Research.

Lockwood, M. (2013). The Political Sustainability of Climate Policy: The Case of the UK Climate Change Act', *Global Environmental Change*, 23(5): 1339–48.

Observatoire des énergies renouvelables (2012). *Worldwide Electricity Production from Renewable Energy Sources*, Stats and Figures Series, 14th inventory, Paris: Observatoire des énergies renouvelables.

REN-21 (2013). *Renewables 2013: Global Status Report*, Paris: Renewable Energy Policy Network for the 21st Century.

Schneider, M., and Froggatt, A. (2013). *The World Nuclear Industry Status Report 2013*, Paris and London: Mycle Schneider Consulting.

Smil, V. (2003). *Energy at the Crossroads: Global Perspectives and Uncertainties*, London: MIT Press.

Smith, A., Stirling, A., and Berkhout, F. (2005). 'The Governance of Sustainable Socio-Technical Transitions', *Research Policy*, 34(10): 1491–1510.

Sovacool, B. K. (2009). 'The Importance of Comprehensiveness in Renewable Electricity and Energy Efficiency Policy', *Energy Policy*, 37: 1529–41.

Turnheim, B. (2012). 'The Destabilisation of Existing Regimes in Socio-Technical Transitions: Theoretical Explorations and In-Depth Case Studies of the British Coal Industry (1913–2010)', Ph.D. thesis, SPRU, University of Sussex.

Turnheim, B., and Geels, F. W. (2013). 'The Destabilisation of Existing Regimes: Confronting a Multi-Dimensional Framework with a Case Study of the British Coal Industry (1913–1967)', *Research Policy*, 42(10): 1749–67.

Unruh, G. C. (2000). 'Understanding Carbon Lock-in', *Energy Policy*, 28(12): 817–30.

Verhees, B. (2012). 'Cultural Legitimacy and Innovation Journeys: A New Perspective Applied to Dutch and British Nuclear Power', Ph.D. thesis, Eindhoven University of Technology.

WBGU (2011). *World in Transition: A Social Contract for Sustainability*, Berlin: German Advisory Council on Global Change.
Winskel, M. (2002). 'When Systems are Overthrown: The "Dash for Gas" in the British Electricity Supply Industry', *Social Studies of Science*, 32(4): 563–98.
Zehner, O. (2012). *Green Illusions: The Dirty Secrets of Clean Energy and the Future of Environmentalism*. London: University of Nebraska Press.

5

The Global Green Economy

Competition or Cooperation between Europe and China?

Hubert Schmitz and Rasmus Lema

5.1. Introduction

Three major waves of change have occurred in the global economy in recent decades. First, there was the reduction of barriers to flows of goods, finance, and information. The second wave involved a shift of economic power from West to East. Third, but only just beginning, is a restructuring towards a greener global economy due to the threat of climate chaos.

China was the main beneficiary of the first wave. Through a process of integration into global value chains, China was able to build up its production capabilities very quickly. This laid the basis for the second wave, the emergence of China as a driver of global change, altering the parameters for competing in the global economy. China has transformed the global map of manufacturing industry. Over a short period it has acquired production capabilities challenging the old industrial powers and forcing them to switch to other stages of the value chain or to other sectors. The green sectors were originally thought be immune from this. During the 2000s Europe expected that its renewable energy industries would retain global pre-eminence and become a bastion of green growth and jobs. Now we know that China is also transforming these green industries.

This chapter examines this challenge from China for Europe's green industry—showing both intense competition and close cooperation. It shows how and why these relationships vary between green sectors and different stages of the value chain. This is the core of the chapter. We then show how the global financial crisis influenced the Chinese challenge for Europe and reflect on how the new division of labour affects the chances of mitigating climate chaos. The chapter thus traces some of the interconnections between two of the challenges discussed in this book: the economic challenge and climate change.

Concentrating on how China challenges Europe in the global green economy might seem a surprising choice given that China has gained a reputation for damaging its own and the world's environment. Indeed, over more than three decades of resource-intensive growth and reliance on coal-fired power, Chinese CO_2-intensity is among the highest in the world (Ecofys, 2014; see also Chapter 1 in this volume). However, having become the world's greatest source of carbon dioxide emissions in the first decade of this millennium, China has now become the world's largest investor in renewable energy and other green sectors (BNEF, 2013). China is thus becoming a major player in the global green economy, with domestic firms rapidly catching up with the world leaders.

The ramifications for European industry vary a great deal, in particular in renewable energy. In hydro-power, China has now become the world's leading exporter of dams for electricity generation. Starting with building up its capabilities by supplying the home market, lead firms such as Sino Hydro have become multinational enterprises. While they operate globally, they mainly compete with European firms in the developing world (Huang and Fang, 2013). In the solar-photovoltaic industry, the local Chinese market only took off more recently. This industry started out as an export-oriented industry and by 'learning from exporting', Chinese producers have moved from supplying components to complete solar panels in a short time-span. By the late 2000s they undercut European (and American) producers, leading to major job losses and prompting a trade war (Fischer, 2012; Luetkenhorst and Pegels, 2014). This painful solar experience has affected not just the European competitors in this sector but the discussion on green growth more generally. As noted by Geels (2013, 83).

> The green growth discourse in Western countries is hindered by cheap Chinese competition, which caused over-capacity, tumbling prices, and bankruptcy or job reductions for many Western wind and solar companies. The industrial shake-out and job losses led politicians to criticize the use of Western government subsidies (for renewable energy) to stimulate Chinese industries.

This chapter sheds light on the assumptions made in the discourse referred to by Geels. It does so by examining the interconnection between China and Europe in the wind power industry. It shows that the changing relationships have to be seen in the context of the global financial crisis. We cannot isolate a 'China effect' because the increasing competitive pressure from China coincides with the global financial crisis.

The European wind power turbine industry was working at full throttle when the financial crisis broke in 2008. Most manufacturers and component suppliers were working at maximum capacity. Due to long lead-times, the effects of the crisis were initially not felt fully; also it was being mitigated to some extent by financial stimulus packages. But 2008 was nevertheless a turning point where overcapacity caused prices to fall. Starting that year, prices fell by more than 20 per cent over the next three years (Gosens and Lu, 2013, 310). By 2014, the European industry was just beginning to rebound from falling prices, major job losses, and dwindling profits.[1]

It is clear that the global financial crisis and the scaling back of national government support has had an adverse effect on the European wind power industry (Luetkenhorst and Pegels, 2014; Sørensen and Springborg, 2010). It is less clear how the rise of China as a major wind power nation has affected the European wind power industry and what the prospects are for competition versus cooperation between European and Chinese firms. This is the question on which we focus in this chapter.

The chapter is structured as follows. Section 5.2 provides essential background information on China's wind power market and its drivers. Section 5.3 analyses the growing competition among the international wind turbine producers in recent years, highlighting the rapid advance of Chinese firms in the global top ten and the parameters on which they compete with European firms. While the competition between European and Chinese lead firms is strong, there are also many examples of cooperation. In order to explain this, section 5.4 sets out the value chain of wind turbine production and innovation. This provides the analytical basis for unravelling surprising forms of cooperation between Chinese and European firms. This cooperation is driven by the pressure on Chinese firms to catch up and improve the quality of their products and the pressure on European firms to gain access to the large and rapidly expanding Chinese market. Reflecting on these findings, section 5.5 asks wider questions about China's role in Europe's crises, stressing the need to differentiate carefully between and within sectors. Section 5.6 discusses economic and political ways forward and section 5.7 sets out questions for further research.

5.2. China and the Green Economy

The purpose of this section is to provide some background on the build-up of China's green economy. China now boasts the world's largest renewable

[1] Tens of thousands of European jobs were lost in the wind power industry during the economic downturn. Denmark has the highest proportion of wind power jobs in Europe. During 2009 when the crisis gathered speed, the Danish wind power industry lost 4,000 industrial jobs out of a total of approximately 28,000 in just one year. Subcontractors accounted for more than 3,000 of those jobs (Sørensen and Springborg, 2010).

energy system, generating over 1 trillion kWh (in 2013) from water, wind, and sun (Mathews and Tan, 2014). After decades of rapid and 'dirty' economic progress, China has begun to set targets devised to reduce the carbon intensity of growth, in particular by investing in renewables.

The Forces of Change

The rationale for investing in low-carbon restructuring differed from European countries, at least initially. While the first push in Europe three to four decades ago was spurred by a growing environmental movement, China's initial turn to green technologies was mainly rooted in industrial development and energy security interests rather than in environmental considerations (Schmitz, 2015). The main concerns were reducing dependence on imported energy, building competitive industries, and (at the local level) creating jobs and raising public revenue (Dai, 2015).

Recent years, however, have seen a growing dissatisfaction with environmental degradation and smog in the rapidly growing Chinese urban areas. Chinese policy-makers are struggling with how to combine high economic growth—still demanded by those left behind in the Chinese take-off—with reducing pollution and carbon dioxide emissions, let alone an absolute decline in the use of fossil fuels (cf. Clark and Hornby, 2014). One part of the solution—the consequences of which are now being felt in many parts of the world—is the fostering of renewable energy industries.

The financial crisis proved to be an important turning point because financial stimuli packages were dedicated to boosting green industries. In 2008 as well as in 2009 financial stimulus packages in China were more ambitious than any other nation when it comes to investments earmarked for green industries (Morrison and Yoshida, 2009). China allocated 5.3 per cent of GDP to these sectors of the economy through green stimulus packages. Countries in the EU also identified investment in renewable energy as a way to overcome the crisis, create jobs, and develop new growth perspectives, but to a much smaller degree. For example, Germany, a forerunner in greening the economy, allocated roughly 0.5 per cent of GDP through the stimulus package in 2008 (Barbier, 2010). As shown later in this chapter, the green stimulus strategy proved to be an inflection point in the growth path in China.

The Significance of China in Renewables

Mathews and Tan (2014) suggest that the year 2013 marked a significant change in the country's energy trajectory because the scales tipped towards electricity generated from renewables compared to coal and nuclear. In 2013, renewables accounted for 60 per cent of *new* electricity generation capacity, of which 11 GW was solar and 16 GW was wind turbines (Clark and Hornby, 2014; GWEC, 2014). This share of added capacity from renewables in electricity generation is higher in China than in most other countries in the world. Although this is still far from the expansion required to meet targets set by climate scientists, Mathews and Tan (2014) assert that carbon dioxide emissions from the power sector in China can be expected to fall soon.[2]

In 2013 the Chinese wind power market was still growing at an annual rate of more than 20 per cent, and China ranked number one globally in terms of cumulative installed capacity with 91.4 GW, or 29 per cent of global capacity (GWEC, 2014; WWEA, 2014). The number one position was achieved at breakneck speed. In 2005, when the domestic industry took off, China's wind power capacity was 1.25 gigawatt (GW), accounting for only 2 per cent of the world market. Between 2005 and 2013, total installed capacity in China thus increased by more than 70 per cent per year. Approximately five years after the 'take-off' in the home market, the Chinese export of wind power equipment had its own take-off in 2011. From a low average of 18 MW/year during the period 2008–10, exports rose rapidly to a level of 692 MW installed capacity in 2013. However, only a very small share of these exports went to Europe, whereas the majority of turbines were exported to emerging markets (Lema et al., 2013, 52–4).

This rapid expansion of wind power capacity in China is mainly a result of favourable government policies. These policies have contributed to rapidly growing internal markets for renewable energy. As in Europe, large subsidies are involved, since the cost of generating electricity from wind and sun is still higher in most Chinese locations than that of fossil-fuel-based energy.

China's rise as a wind turbine producer is thus closely related to the size and growth of its wind energy market and the policies that support it. The global demand for wind energy is shifting to the East. European countries and the US have long been the dominant markets in the global wind power sector but by 2010, more than half of newly installed wind power capacity was added outside these traditional markets. China has been the main driver of this development. Within just half a decade or so, China has managed to build the largest market for wind power in the world. This is the home market from which Chinese firms are now taking off and competing internationally.

[2] The current high share of coal in the energy mix, combined with rapid economic growth in heavy industries, has put China in the position of being the biggest CO_2 emitter globally (IEA 2014b). However, a large share of China's emissions is due to exports—approximately 23% of Chinese emissions are due to China's insertion into global value chains catering for the European and North American markets (Watson and Wang, 2007).

5.3. Competition

The world keeps observing China's economic rise with a mixture of admiration and trepidation: admiration because the speed and depth of the country's economic transformation are unprecedented in history; trepidation because China has increased the competitive pressure on industry in many countries and sectors. There is now a substantial literature on the impact of China's industrial transformation in creating both considerable threats but also new opportunities for industry in other countries (Eichengreen, 2006; Kaplinsky and Messner, 2008; Jenkins and Barbosa, 2012; Fujita, 2013). This chapter seeks to show how this balance of threats and opportunities has been unfolding in the European wind turbine industry. This section discusses the increasing competition between Europe and China while section 5.4 examines the opportunities for cooperation.

The Rise of Chinese Lead Firms

The rotating blades of wind turbines have become an iconic symbol of the green economy. Until recently, Europe was the global leader in producing and using such wind turbines. This European leadership rested largely on the green investments of two countries, Denmark and Germany. Most of our discussion of Europe refers to these two countries. Denmark was the pioneer in producing and using wind power, but Germany caught up rapidly. Today, both countries remain global leaders in innovation but China has become number one in the production of wind turbines. China is also catching up in developing new wind power technology (Lewis, 2013). Unpacking and understanding this overall picture is essential for the central argument of this chapter concerning the coexistence of competition and cooperation.

Competition is intense and most visible when we concentrate on the lead firms of the wind turbine industry. Table 5.1 gives some idea of the competitive pressure on Europe and the speed of China's advance. It shows the top ten turbine manufacturers in the world. While in 2005, seven of the top ten firms were European and none were Chinese, by 2010 there were four Chinese firms in the top ten. By 2013 this number had reduced to three, but four of the following five wind turbine manufacturers just below the top were also Chinese (NAW, 2014). Europe remained strong with five in the top ten by 2013 but clearly less strong than in 2005 (Table 5.1). The big loser has been General Electric, whose market share declined from 17.7 per cent in 2005 to less than 5 per cent in 2013.

The lead firms compete head-on, in particular for the major contracts—equipping the new wind farms. The market is differentiated according to different types and sizes of turbines (customized for different wind speeds,

Table 5.1. Top ten global wind turbine manufacturers 2005–13

2005			2010			2013		
Company	Share	Origin	Company	Share	Origin	Company	Share	Origin
1. Vestas	27.90%	EU	1. Vestas	14.8%	EU	1. Vestas	13.2%	EU
2. General Electric	17.70%	US	2. Sinovel	11.1%	CN	2. Goldwind	10.3%	CN
3. Enercon	13.20%	EU	3. General Electric	9.6%	US	3. Enercon	10.1%	EU
4. Gamesa	12.90%	EU	4. Goldwind	9.5%	CN	4. Siemens	8.0%	EU
5. Suzlon	6.10%	IN	5. Enercon	7.2%	EU	5. Suzlon	6.3%	IN
6. Siemens	5.50%	EU	6. Suzlon	6.9%	IN	6. General Electric	4.9%	US
7. Repower	3.10%	EU	7. Dongfang Electric	6.7%	CN	7. Gamesa	4.6%	EU
8. Nordex	2.60%	EU	8. Gamesa	6.6%	EU	8. United Power	3.9%	CN
9. Ecotécnia	2.10%	EU	9. Siemens WP	5.9%	EU	9. Mingyang	3.7%	CN
10. Mitsubishi	2.00%	JP	10. United power	4.2%	CN	10. Nordex	3.4%	EU
Others	5.00%		Others	17.5%		Others	31.6%	

Note: World market shares
Source: BTM (2006, 2011, 2014)

temperatures, and onshore/offshore installation). Overall, however, Europe has the edge with regard to product differentiation, large-scale and offshore turbines, and reliability, and China with respect to price and medium- and small-scale onshore turbines.

The Sources of Economic Power

In order to understand the competitive dynamics, it helps to observe how China was able to increase its share of the global market. Here one can distinguish four sources of competitiveness which need to be distinguished individually but which also significantly reinforce each other.[3] The first is *the strength of the home market*. The Chinese Government—understandably concerned with energy security—has fostered the production of renewable energy. The Renewable Energy Law of 2005 was the central piece of legislation. Along with other complementary policies, it helped to create a rapidly expanding internal market for wind power. Foreign enterprises were not prevented from competing in this market but Chinese enterprises were favoured, receiving government support through various means, some visible

[3] In this chapter we do not deal with a fifth source coordination power, one which derives from highly modular supply chains in China (Lema et al., 2013).

(e.g. local content requirements in place during 2005–9), others less visible (e.g. bidding for state-funded projects). Since the Chinese market was large and fast-growing, success in this market had a major impact on global market shares, which are summarized in Table 5.1.

The second source of Chinese competitiveness is *producer power*. The large size and rapid growth of the Chinese market enabled Chinese turbine manufacturers to adopt a model of industrial organization geared towards economies of scale. A turbine is a complex product typically consisting of over 10,000 parts. While leading European firms—such as Vestas of Denmark and particularly Enercon of Germany—produced many of these parts in-house (seeking to constantly improve design and quality), their Chinese counterparts relied much more on buying components from suppliers who also supplied other turbine makers and were thus able to achieve economies of scale and reduce costs.

The third source of competitiveness is *financing power*. This is as yet little explored in the literature but is of increasing importance. The essential point is that Chinese firms can offer supplier credit, but this is much more difficult for Western firms. This matters greatly for their customers since investment requirements for wind farms are high and time-frames are long. Project finance may become particularly important for competing in export markets. Compared with their Western counterparts, Chinese companies have deep financial pockets. Sinovel, for example, has a US$6.5 billion line of credit from government-owned banks (Zeller and Bradsher, 2010). China Exim Bank has injected capital into Goldwind and Ming Yang to support foreign expansion. Such support opens up the possibility of an export model that has not been directly utilized by European firms—the twinning of wind farm project finance and turbine exports. Many new projects undertaken abroad by Chinese turbine firms have been implemented with tag-along finance. The finance effect is beginning to show. 'Competition from Chinese manufacturers has dragged down prices for wind turbines, while Europe's economic slowdown has tightened credit markets, curbing funds for project development' (BNEF, 2012a).

A potential fourth source of competitiveness is *innovation power*. There is controversy over how real this is. Some argue that Chinese firms rely largely on technologies licensed from European firms. This view is supported by patent statistics which reveal that China occupied an insignificant position in wind energy patenting well into the first decade of this millennium (McDowall et al., 2013). Others argue that Chinese firms have moved on and started to accumulate significant innovation capabilities. Since the locus of innovation power is essential for the future of Europe's industry, the next section delves deeper into this issue, unpacking the parts of the value chain concerned with innovation. This then enables us to gain new insights into the nature of the relationships between European and Chinese firms.

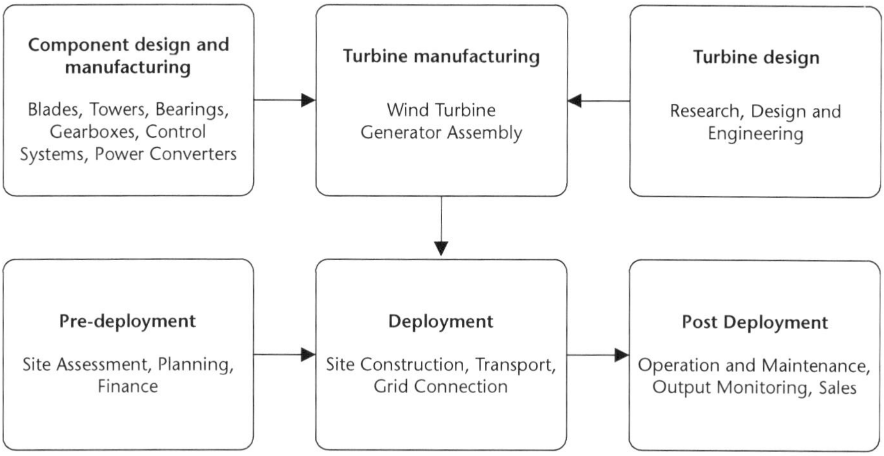

Figure 5.1. The wind power value chain

The head-on competition between Europe and China which we have stressed so far—and which dominates the discourse in the media and business press—occurs between the lead firms in turbine manufacturing. A different picture emerges, however, if we shift our focus to different stages of the value chain, namely turbine design, component design and deployment (Figure 5.1). The next section examines these different parts of the value chain.

5.4. Cooperation

Value chain analysis offers a valuable analytical tool for understanding how the global map of industrial production has changed (Gereffi, 1999). It has proved particularly valuable for understanding the speed with which newcomers come to compete in the global economy (Schmitz, 2006) and how different ways of organizing the global value chain affect the ability to compete (Sturgeon, 2002). The innovation part of the value chain, however, has remained relatively underexplored.

This is beginning to be addressed in new work on the organizational decomposition of the innovation process—ODIP (Schmitz and Strambach, 2009). The essence of value chain analysis, namely slicing the value-adding process into separate activities, was extended to the innovation process. The starting point for this line of investigation was the fundamental change in the way innovation is organized. Innovation activities that used to be carried out in-house are now often transferred to independent suppliers of knowledge-intensive business services (KIBS) or suppliers of key components. This helps with understanding the changing global innovation map. Using such ODIP analysis, we have shown shifts in innovation power to Brazil and India, as evidenced in the auto and software industries (Lema et al., 2012). Analysing the organizational decomposition of the innovation process is essential for understanding the issue that is central to this chapter—the relationships between European and Chinese firms in the wind power industry.

Cooperation in Licensing and Co-Design Relationships

European wind turbine makers rely more on their internal capabilities than their Chinese counterparts, but even they cannot do everything in-house. As noted earlier, a wind turbine is a very complex product, with some parts requiring highly specialized expertise. A good part of this expertise resides in knowledge-intensive business services. These KIBS firms do not have a manufacturing capacity but instead adopt a business model focused on selling R&D and consultancy services and licences for wind turbine design. To give some examples, Windtec of Austria was among the first to sell licences in the market, becoming a technology supplier to Germany's Fuhrländer and other European firms. Norwin of Denmark focused on turbine design and entered into an agreement with the German firm Preussac, which then manufactured turbines based on the resulting licence agreement. Some of these KIBS emerged out of the earlier consolidation and restructuring of the European wind power industry. For example, Norwin was founded by three engineers from the R&D department of Danwind, a wind turbine firm which went out of business in 1991. Such KIBS play an important role with regard to wind power innovation in Europe. They are also important for the build-up of production and innovation capability in China.

When China entered the wind power industry, it had no turbine technology of its own. It relied almost entirely on licensing technology from Western firms. Such licences were acquired not from the lead firms but from smaller turbine makers and in particular from wind KIBS (Lewis, 2013, pp. 136-37). These KIBS firms have developed a revenue stream based on licence fees from their customers abroad, in particular in China. The Chinese manufacturers used these licences to build up their own production capabilities. All this is relatively well known. What is less well known is that some of the Chinese manufacturers have also now begun to build up their own innovation capabilities. This is being done through various learning channels. One of the most important channels is the relationship with the European KIBS: what was originally a licensing relationship has often become a co-design relationship. Turbines or components are designed jointly between a Chinese manufacturer and a European design house. The Chinese side gains access to top European engineers and the European side gains (indirect) access to the Chinese market.

The underlying shift in technology relationships is captured by Lewis (2013) in her observations on one of the China's lead firms: 'Goldwind, like many other wind companies, began its entry into the wind industry by acquiring a technology licence from a foreign company. As it became more experienced, it sought out a German design firm to work with rather than a manufacturer that could also be a competitor' (2013, 133). This case is not an exception, as shown in Figure 5.2, which depicts the relationships between twenty-six Chinese turbine makers and eighteen (mainly) European KIBS. While not claiming to be comprehensive, the figure indicates that such relationships are now fairly common. Some are mainly about licensing, others about co-design, while some include both. It is not yet clear how durable the arrangements of cooperation in design are, but clearly the inter-firm relationships between China and Europe are changing over time, helping China to break through from production to innovation.[4]

While the globalization of the innovation process has contributed to this process, it was also supported by favourable domestic factors, notably public policies (see section 5.2; Cunningham, 2014), investments in local innovation systems (Gosens and Lu, 2013), investments in firm-level learning (Qiu and Anadon, 2012) and other factors typically emphasized in the catch-up literature (Lee, 2005; Bell and Figueiredo, 2012).

Cooperation across Value Chain Functions

In addition to the collaborative relationships between European turbine designers and Chinese turbine manufacturers, many other instances of cooperation along the value chain can be observed. For example, European component designers and component manufacturers are working with Chinese firms. While European turbine manufacturers generally face a tough time in China, certain suppliers of key components are working for, rather than competing with, Chinese firms. Control systems software is one of the areas in which European firms have a competitive edge and Chinese firms are seeking to learn from them. For example, the Danish firm Mita-Teknik works with Sinovel on control system software. Sinovel and Mita-Teknik had originally entered into an agreement under which Sinovel purchased controls systems and software from Mita-Teknik, but under a new agreement the two companies will develop the next generation of control systems jointly (Lema, 2012).

European firms are also helping China to develop its offshore wind power base. Siemens was responsible for the Jiangsu Rudong Offshore farms, which

[4] The shift from licensing to co-design has also received increasing attention from Chinese scholars, notably Qiu and Anadon (2012), Ru et al. (2012), Wang et al. (2012), Zhou et al. (2012), and Dai et al. (2015).

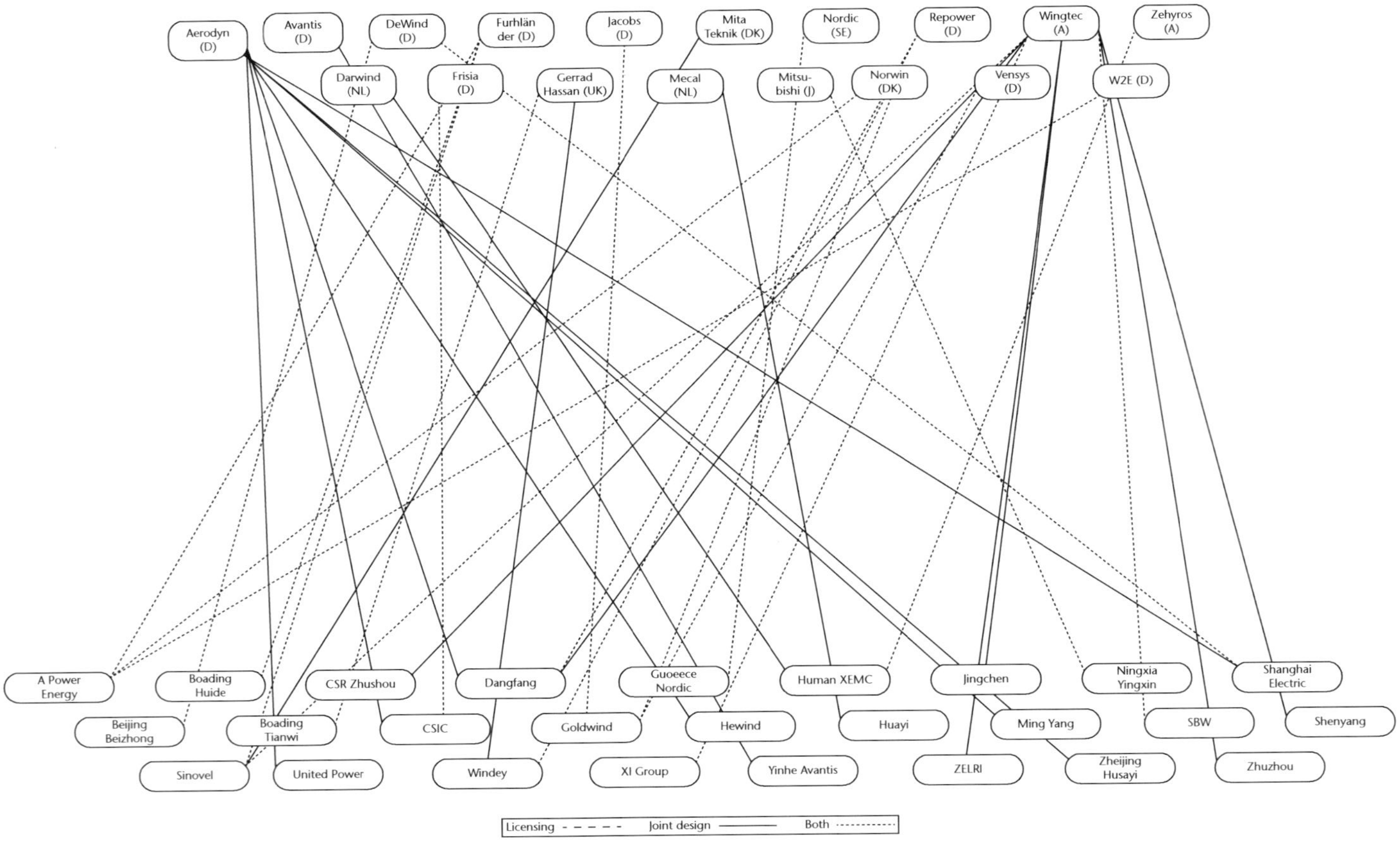

Figure 5.2. Links between European KIBS and Chinese manufacturers

were connected to the grid in May 2012. Besides developing the wind farm, Siemens also entered into a joint venture with Shanghai Electric, which was established to focus on offshore projects. Such projects bring European expertise to the demanding offshore segment of the wind power industry, which is still relatively underdeveloped in China. On the other hand, China has capabilities for producing at low cost. Some European utilities are pondering how to use turbine foundations produced in China to bring down the costs of offshore projects.

The transformation of China–Europe relationships does not stop here. Lead firms from both sides have become global firms rather than remaining purely European or purely Chinese firms. Vestas has established a sizeable R&D laboratory in China and Goldwind has acquired majority control of the German wind turbine maker, Vensys. Titan Wind Energy acquired a tower factory for Vestas of Denmark, thus rescuing jobs that would otherwise have been lost due to the major restructuring process undertaken by Vestas that was triggered by the financial-economic crisis. Merely contrasting European and Chinese industry risks missing out the transformation of relationships between them that is under way.

To sum up, section 5.3 revealed fierce competition between European and Chinese lead firms of the wind power industry. However, this section has shown that this fierce competition coexists with cooperation between Chinese and European firms. The relationships of cooperation become apparent once we closely examine the value chain, in particular the innovation part. Unpacking the innovation process and identifying changes over time are essential for bringing to light the evolving pattern of inter-firm cooperation across the Europe–China divide.

5.5. China's Rise—Europe's Crisis?

The picture which dominates the media is that China's rise has contributed to Europe's crisis. Careful research shows that the China effect varies between countries and sectors (Venables and Yueh, 2006; Jacques, 2012). Nevertheless, competitive pressures have increased for most industrial enterprises, including those operating in the green economy.

The reconfiguration of competition and cooperation has important ramifications both from a European economic crisis perspective and from a sustainability perspective. As shown in this chapter, the assertion that cheap Chinese competition is the root cause of overcapacity, unemployment, and falling share prices in the wind power industry in Europe is an exaggeration. In wind power, the competition from China on European soil is still very limited (Lema et al., 2013, 52–3). The mitigating effects include high transportation costs, the inability (of Chinese lead firms) to meet European quality standards, and certain tendencies to favour national firms in major public projects (Lema et al., 2011).[5]

Overcapacity in Europe arose first and foremost from austerity policies which scaled back government support for wind power and led to the cancellation of major projects. For example, Spain completely removed its feed-in tariffs for renewable energy projects in 2012, while many countries reduced their tariffs (REN21, 2012). The offshore sector was also affected. Banks lowered their offshore project funds (Wieczorek et al., 2013, 3015) and governments implemented cancellations and delays. For example, in 2014 Denmark put the 600MW Kriegers Flak, the world's first project to supply three countries (Denmark, Sweden, and Germany), on hold (Jensen 2014).

Figure 5.3 shows how the market contracted in Europe (and continued growing in China) when the global financial crisis hit in 2008. It was this contracting which affected the industry adversely. Wind was not alone; the

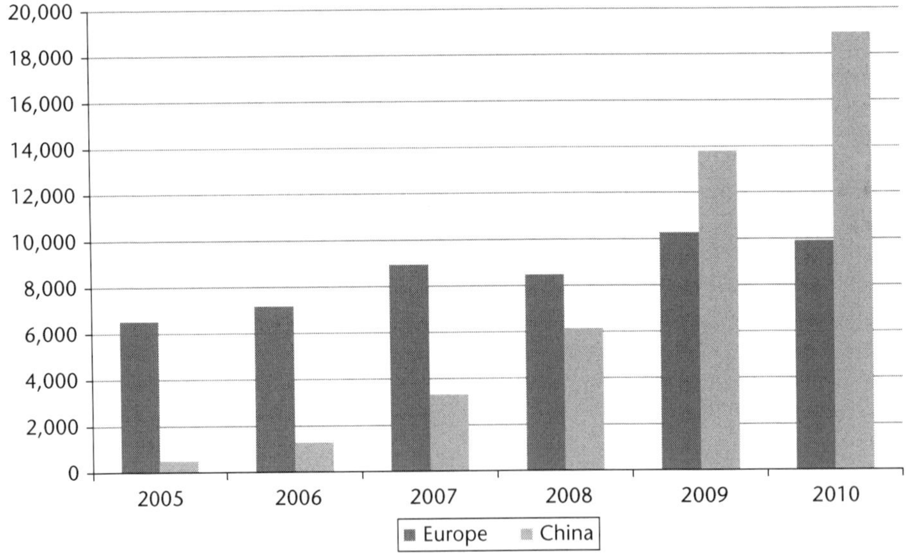

Figure 5.3. Annual wind power installation in EU and China, 2005–10 (MW)

[5] Many countries have implemented formal local content requirements (Kuntze and Moerenhout, 2013). As Fitch-Roy (2013, 51) puts it: 'Ministers of trade are adopting a pragmatic attitude to local content rules that reflects the trade-offs faced by an intrinsically political business.' In Europe 'protectionist' tendencies are more hidden within tendering decision-making processes but arguably account for the fact that domestic or nationally incorporated firms dominate the markets in countries such as Denmark, Germany, and Spain.

entire renewable energy sector was hit severely. At the time, the consequences were assessed as follows:

> The investment surge of recent years was just starting to ease the supply-chain bottlenecks when the credit crunch arrived and cut demand. The result has been a dramatic and permanent change to the dynamics of the industry. On the supply side prices are falling towards marginal costs, and several players will consolidate. On the demand side renewable energy targets will still drive utilities to build projects, but fewer developers and independent power producers will be involved. (Fritz-Morgenthal et al., 2009, 1)

The picture which thus emerges is that, while China is driving major changes in the wind power industry, it would be misleading to see it as the main cause of the crisis in this industry. China's rise coincided with the global financial-economic crisis, but Chinese competition was not a major cause of job losses and dwindling share prices in Europe. Subsidies for wind energy in Europe accrue almost exclusively to European firms as Chinese firms are yet to make a major foray into the European market. However, the European firms are aware that this might change, since they compete with Chinese firms in third markets and know their cost advantages. They are also forewarned by the rapidly changing fortunes of the European (and American) solar power industries.

Indeed, it is worth reflecting on our findings in a comparative way. The balance between competition and cooperation in Europe–China relationships differs between green industries, notably between wind and solar-photovoltaic industries. A brief comparison helps to see the need for such a differentiated analysis.

As shown in this chapter, China's entry into the wind energy industry is reshaping the relationships between Europe and China in major ways. A big domestic market was critical for building China's production capacity; this in turn put pressure on European incumbents and drove down costs.

China has also caused major disruption and driven down costs in the solar panel industry, but the dynamics were different. Here Chinese firms managed to build up producer power by catering for the European market; initially the domestic market—and reducing green house gas emissions domestically—played less of a role. Figure 5.4 shows how solar-PV production was driven by exports during industry 'take-off'.

Exports enabled China to emerge as the new leader (replacing the EU) in producing solar-PV equipment (Fischer, 2012). However, in contrast to the wind sector, in which Chinese manufacturers still produce predominantly for the Chinese market, the Chinese PV sector has grown mainly on the basis of policy support for solar energy deployment in Europe. The development of China's PV sector originated from the production of PV cells and modules. By focusing on these elements, enterprises concentrated on the steps

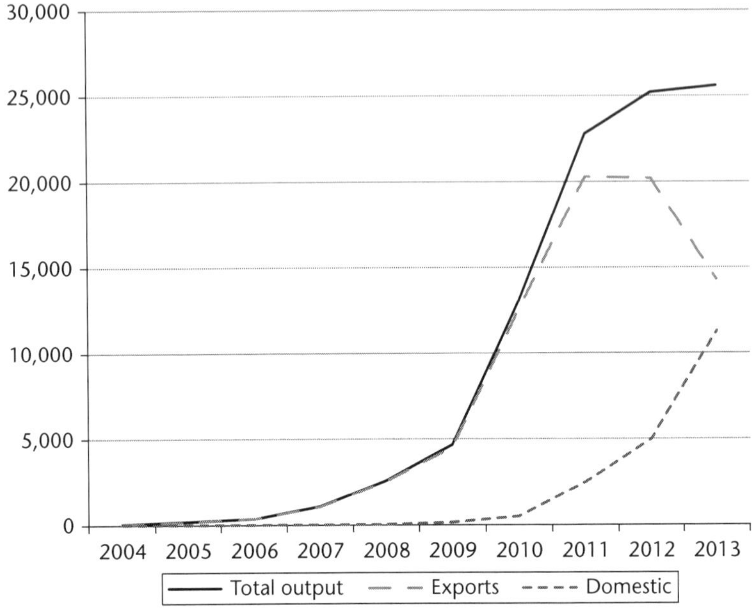

Figure 5.4. Chinese sales of solar-PV cells: exports and domestic market (megawatt)

in the PV value chain where they had competitive advantages due to low labour costs, economies of scale, and comparatively weak environmental standards applied to production processes. The development of the Chinese PV sector and exports to Europe have been facilitated by modularization and comparatively low transportation costs.

In China, government support for these activities was driven by local economic considerations and was similar to support given by local government to other export-oriented industries (Iizuka, 2014). Support schemes for the use and deployment of PV energy technology in China evolved in the context of the global financial crisis when export-oriented producers of PV cells and modules faced falling external orders, particularly in Germany. Hence, when the central government started to support PV energy use within China in 2009, the Chinese PV industry was already a highly capable producer and a fierce competitor in global markets. Figure 5.4 shows how domestic sales were increased dramatically when exports began to stagnate in 2011 and then fell from 2012—thus compensating for decreasing external demand and ensuring overall growth in output.

This was almost the complete opposite to the route taken in the wind turbine industry where the local Chinese market was the mainstay for domestic firms and where exporting came later. The patterns of competition and collaboration therefore differed in important ways both during and after the financial crisis.

In summary, the two green sectors are now highly globalized and Europe's leadership is challenged in both. However, keeping in mind our focus on competition and cooperation with China, the value chains lens shows that the challenges take different forms in the two sectors. Hence there is also a need for differentiated responses when it comes to finding ways forward.

5.6. Ways Forward

There is no shortage of contentious accounts of EU–China relationships in renewable energy. China is often portrayed as causing problems for the green industry and the climate. This chapter highlights the need to go beyond media headlines and simplistic rhetoric. Cutting through the complexities is essential for understanding the rapidly changing world.

Currently, we observe that competition between the lead firms of the wind and solar-PV industries is increasing rapidly, not least as a result of China's increasing production and innovation power. From a green transformation perspective, this development is to be welcomed as it drives down the price of renewable-energy technology, supporting the reduction of greenhouse gases. Policy-makers in China and Europe, of course, are concerned not only about the global gains, but also the distribution of those gains among nations.

In a rapidly globalizing renewable-energy sector, however, it is becoming increasingly difficult to define the 'national content' of a wind turbine or a solar-PV panel. Production and innovation in both sectors are increasingly globalized. This can be turned to mutual advantage. For example, the Chinese system of renewable-energy production is entering a new stage where it is increasingly important to improve the quality and reliability of technology. Hence, Chinese lead firms should seek to collaborate with European companies at various stages of the value chain to develop this expertise. In turn, Europe, in its current stage of getting back on its feet after a prolonged financial and economic crisis, is in need of more affordable renewable-energy technology. European lead firms, under pressure to reduce costs, may be able to cooperate with Chinese firms and access capabilities and resources for scaling up.

The solar-PV manufacturing industry was severely affected by Chinese competition and 'rescue missions' seeking to breathe life into European solar-PV manufacturing are probably futile. Rather, policy-makers should seek to support those elements of the value chain that are globally immobile. Distinguishing between the manufacturing chain and the deployment chain is critical. While it is true that European lead firms of the production value chain are losing market shares due to increased competition from Chinese lead firms, European installation, operation, and maintenance companies in the downstream segment of the value chain are likely to benefit from the expansion of the market as a result of falling prices for solar-PV equipment. Policies designed to support the solar deployment chain while drawing on Chinese manufacturing capabilities may contribute to job creation as well as green transformation.

In wind energy, policy-makers may seek to support collaborative relationships that benefit European industry, Chinese industry, and the global climate. Chinese involvement in select value chain steps, such as manufacturing of foundations for offshore deployment or the provision of bearings, may help to lower costs significantly. More ambitiously, cross-continental consortia may be formed between actors such as banks/institutional investors, utilities, manufacturers, and operators for projects in Europe, China, and the rest of the world. Pushing down costs and pushing up scales requires experimenting with different approaches.

The China effect has certainly been painful for German and other affected producers in the solar-PV industry. The question is whether a similar process may occur in wind power. The gloomy European observer might suggest that, while the many relationships between European design houses and Chinese wind power producers may create or preserve some jobs and profits in Europe in the immediate term, the industry needs to be viewed from a dynamic perspective: these design houses and their engagement in co-design relations may be undermining the competitiveness of European producers in the long run, with the consequence that the wind industry in Europe will follow a similar pattern to the solar-PV industry. We do not think this likely. Due to the markedly different technological characteristics of solar-PV and wind energy and the associated differences in industrial organization, there will be a less dramatic China effect in the manufacturing part of the wind power value chain. It is more likely that regional production networks will be strengthened. As with the automotive industry, 'regional' European, Asian, and American networks are likely to coexist and different lead firms will dominate and compete globally while producing and sourcing mainly locally.

Value chain analysis can be a powerful tool for designing intelligent policies for the support of European renewable energy sectors that are faced with global competition in different ways. The multiple challenges confronting Europe today (climate change, reduced public finances, and competitive pressures from Asia) call for highly selective and directed strategies to support markets and new collaborative business models.

Developing and implementing such selective green industrial policy is not easy. The technological, economic, and political uncertainties are considerable (Schmitz et al., 2015). In order to succeed, green industrial policy needs to focus on specific subsectors and adopt specific instruments aimed at developing and deploying specific technologies. Predicting which technology to support for how long means making choices about allocating policy rents on the basis of highly incomplete information. The uncertainties are exacerbated by the long time-horizons needed for new energy systems to become viable. Yet mitigating climate change requires action now. Climate and earth scientists warn that delaying action means incurring rising costs and potentially irreversible damage.

This combination of uncertainties makes green industrial policy particularly challenging. Elsewhere we have argued that the way forward lies in putting politics centre stage and taking four critical steps. The first is to recognize that no single actor has the resources necessary to bring about the green transformation. The second is to recognize that, within government, civil society, and business, there are actors seeking to block or slow down the green transformation. Third, attention needs to focus on supportive alliances across these categories. Fourth, including actors with different motives helps one to understand and accelerate the green transformation process (Schmitz, 2015).

The last point is particularly critical for the concerns of this book. There is a range of actors who can support climate-relevant policies (e.g. by investing, providing expertise, or lobbying) but their motive need not be mitigating climate change; the main motive might be to secure energy, to build competitive green industries, or to foster green jobs—with climate change mitigation at best a 'co-benefit' (Dubash et al., 2013). Acting upon this can be a 'game changer' and provide the basis for building transformative alliances. The problem for European countries is that China's rise makes it more difficult to build such alliances. Mobilizing political support for green policies in Germany, Denmark, or other European countries is much harder if many of the resulting jobs are created in China. So the question at the core of this chapter—competition or cooperation between Europe and China—has ramifications for the political dynamics behind green industrial policies.

5.7. Questions for Future Research

The insights provided in this chapter contribute to understanding a number of questions which are not the core focus here but are nevertheless critical for Europe and the rest of the world. We set out five questions that can provide a starting point for future research.

First, the chapter indicates that business relations between Europe and China are characterized by cooperative competition in important areas of the value chain. The changes over time have implications for jobs, taxes, and long-run competitiveness. The changes have often been painful for European lead firms and their workers. For example, while collaborative relationships between KIBS providers in Europe and turbine manufacturers in China are mutually beneficial to the parties involved, such relationships may be against the interest of European industry as a whole because they may undermine long-run competiveness, jobs, and taxable income. Future research needs to map out more precisely the strategic implications and the distribution of costs and benefits emerging out of those changing relationships.

Second, there is worldwide concern that China's economic rise has triggered a global 'race to the bottom'. While the mobilization of China's industrial reserve army has exerted a downward pressure on wages and prices in many industries, its firms are not stopping at competing by price cutting but are also now competing by innovating (Altenburg et al., 2008). The wind power industry analysed in this chapter shows that the country is undergoing this transformation in record time. New research is needed that gives specific attention to the time scale over which China is catching up in this technological area. The need to pay greater attention to the time-scales of building up innovation capability has been set out in detail by Bell (2006). This chapter suggests that such research should focus on the combination of internal and external capabilities. Collaboration with external KIBS has provided a relatively fast route to catching up, but exactly how fast has it been and how deep do the Chinese innovation capabilities go?

Third, there is worldwide concern that China's rapid growth is polluting the planet and harming the climate. While China has indeed contributed more than any other nation to the recent increases in carbon dioxide emissions, it has also become the number one investor in renewable energy (BNEF, 2013). This chapter suggests that these investments are being used productively to establish competitive green industries. Mathews (2012) even suggests that China may be on its way to developing a 'green model of industrial capitalism'. Examining the reality of this proposition requires research which compares the growth rates of investment in low versus high carbon industries and which compares the strength of alliances supporting low versus high carbon industries. Such studies need to take into account the insight that 'green alliances' are not necessarily driven by environmental concerns but also by concerns with energy security and competitiveness in green industries.

Fourth, it is clear that global benefits are emerging because China's entry into the wind power industry is lowering the costs of renewable energy technology. Whether the lower equipment costs will speed up the deployment of wind turbines in Europe and other parts of the world is, however, not yet clear. Other considerations influence the deployment of wind technology, in particular reliability and energy output. Lifetime costs and actual electricity generation capacity are more important than upfront costs and nominal capacity. This depends on the demand preferences of different markets. The big unknown is the preference in new markets. A recent report suggests that future growth in the wind power industry will be driven more by new markets such as South Africa, Chile, and Russia (MAKE, 2013). How effective will Chinese firms be in bringing down the cost of energy and opening up new markets?

Fifth, China is building up its own innovation capabilities. The question is whether innovations coming out of China will differ substantially from innovations coming out of Europe and, if so, how and why. Are different specializations emerging that may give rise to a new division of labour? It would be particularly interesting to examine whether solutions from China are more affordable and appropriate for low and middle income countries, enabling a closer alignment of environmental and development objectives. Examining whether this is occurring entails looking beyond core technologies and examining the relevance of European versus Chinese business models for poorer parts of the world.

References

Altenburg, T., Schmitz, H., and Stamm, A. (2008). 'Breakthrough? China's and India's Transition from Production to Innovation', *World Development,* 36(2): 325–34.

Barbier, E. B. (2010). 'How is the Global Green New Deal Going?', *Nature,* 464: 832–3.

Bell, M. (2006). 'Time and Technological Learning in Industrialising Countries: How Long Does it Take? How Fast is it Moving (If at All)?', *International Journal of Technology Management,* 36(1–3): 25–39.

Bell, M., and Figueiredo, P. N. (2012). 'Innovation Capability Building and Learning Mechanisms in Latecomer Firms: Recent Empirical Contributions and Implications for Research', *Canadian Journal of Development Studies,* 33(1): 14–40.

BNEF (2012). *Global Trends in New Energy Investment 2012,* Frankfurt: Frankfurt School of Finance and Management.

BNEF (2013). *Global Trends in New Energy Investment 2013,* Frankfurt: Frankfurt School of Finance and Management.

BTM (2006). *World Market Update 2005,* Ringkøbing: BTM Consult.

BTM (2008). *World Market Update 2007,* Ringkøbing: BTM Consult.

CCB International (2011). *Alternative Energy: China Wind Power,* Hong Kong: China Construction Bank Corporation.

Clark, P., and Hornby, L. (2014). 'A Shift in the Wind (Chinese Pollution)', *Financial Times,* 28 Apr., <www.ft.com/cms/s/0/9ea031fe-cc5c-11e3-9b5f-00144feabdc0.html#axzz3WzYAPgkK>.

Cunningham, E. A. (2014). 'Green Innovation in China: China's Wind Power Industry and the Global Transition to a Low-Carbon Economy', *Review of Policy Research,* 31(2): 150–2.

Dai, Y. (2015). 'Who drives climate-relevant policy implementation in China?' , IDS Evidence Report 134, Brighton: Institute of Development Studies at the University of Sussex.

Dai, Y., Zhou, Y., Xia, D., Ding, M., and Xue, L. (2014). *Innovation Paths in the Chinese Wind Power Industry*, DIE Discussion Paper, Bonn: German Development Institute, forthcoming.

Dubash, N. K., Raghunandan, D., Sant, G., and Sreenivas, A. (2013). 'Indian Climate Change Policy: Exploring a Co-Benefits Based Approach', *Economic and Political Weekly*, 48(22): 47–61.

Earth Policy Institute (2014). *Data Highlights: China's Solar Panel Production to Double by 2017*, Data Highlight, 47, Washington, DC: Earth Policy Institute.

Ecofys (2014). 'International Comparison of Fossil Power Efficiency and CO_2 Intensity', Update, Sept. 2014, Utrecht.

Eichengreen, B. (2006). 'China, Asia and the World Economy: The Implications of an Emerging Asian Core and Periphery', *China and World Economy*, 14(3): 1–18.

EWEA (2014). *Wind in Power: 2013 European Statistics*, Brussels: European Wind Energy Association.

Fischer, D. (2012). 'Challenges of Low Carbon Technology Diffusion: Insights from Shifts in China's Photovoltaic Industry Development', *Innovation and Development*, 2 (1): 131–46.

Fitch-Roy, O. (2013). 'Local-Content Rules have their Role in an Imperfect Market', *Recharge*, 4(13): 51.

Fujita, M. (2013). *Does China's Economic Rise Help or Hinder the Development of its Neighbours?*, Evidence Report 45, Brighton: IDS.

Geels, F. (2013). 'The Impact of the Financial-Economic Crisis on Sustainability Transitions: Financial Investment, Governance and Public Discourse', *Environmental Innovation and Societal Transition*, 6: 67–95.

Gereffi, G. (1999). 'International Trade and Industrial Upgrading in the Apparel Commodity Chain', *Journal of International Economics*, 48(1): 37–70.

Gosens, J., and Lu, Y. L. (2013). 'From Lagging to Leading? Technological Innovation Systems in Emerging Economies and the Case of Chinese Wind Power', *Energy Policy*, 60: 234–50.

GWEC (2014). 'Global Statistics', Global Wind Energy Council, www.gwec.net (20140429).

Huang, B., and Fang, D. (2013). 'Internationalization of Chinese State-Owned Construction Enterprises: A Case Study of Sinohydro', *Journal of Engineering Management*, 3: 111–16.

IEA (2012). *National Survey Report of PV Power Applications in China*, ed. Xu, Honghua, et al. Paris: International Energy Agency Photovoltaic Power Systems Programme.

IEA (2013). 'Energy Statistics', <http://www.iea.org/statistics>.

IEA (2014a). *Snapshot of Global PV 1992–2013*, Paris: International Energy Agency Photovoltaic Power Systems Programme.

IEA (2014b). *World Energy Investment Outlook*, Paris: International Energy Agency.

Iizuka, Michiko (2014). *Diverse and Uneven Pathways towards Transition to Low Carbon Development: The Case of Diffusion of Solar PV Technology in China*, MERIT Working Papers, 002, Maastricht: United Nations University—Maastricht Economic and Social Research Institute on Innovation and Technology (MERIT).

Jacques, M. (2012). *When China Rules the World*, 2nd edn, London: Penguin.

Jenkins, R., and Barbosa, A. D. F. (2012). 'Fear for Manufacturing? China and the Future of Industry in Brazil and Latin America', *China Quarterly,* 209: 59–81.

Jensen, Karin (2014). 'Danes Bump Kriegers Flak to 2022', *Recharge News,* <http://www.rechargenews.com/wind/article1366276.ece>.

Junfeng, L. (2012). *China Wind Energy Outlook 2012,* Beijing: Chinese Wind Energy Association.

Kaplinsky, R., and Messner, D. (2008). 'The Impact of Asian Drivers on the Developing World', *World Development,* 36(2): 197–209.

Kuntze, J. C., and Moerenhout, T. (2013). *Local Content Requirements and the Renewable Energy Industry: A Good Match?,* Geneva: International Centre for Trade and Sustainable Development.

Lee, K. (2005). 'Catch-up: Barriers and Opportunities', *Asian Journal of Technology Innovation,* 13(2): 97–131.

Lema, R. (2012). 'The Decomposition of Innovation and China's Catch-up in Wind Power Technology', paper presented at the 10th Globelics Annual Conference, Hangzhou, China, Novr.

Lema, R., Berger, A., Schmitz, H., and Song, H. (2011). 'Competition and Cooperation between Europe and China in the Wind Power Sector', *IDS Working Papers,* 2011: 1–45.

Lema, R., Berger, A., and Schmitz, H. (2013). 'China's Impact on the Global Wind Power Industry', *Journal of Current Chinese Affairs,* 1/2013: 37–69.

Lema, R., Quadros, R., and Schmitz, H. (2012). *Shifts in Innovation Power to Brazil and India: Insights from the Auto and Software Industries,* Research Report, 73, Brighton: Institute of Development Studies.

Lewis, J. I. (2013). *Green Innovation in China: China's Wind Power Industry and the Global Transition to a Low Carbon Economy,* New York: Columbia University Press.

Luetkenhorst, W., and Pegels, A. (2014). *Stable Policies—Turbulent Markets, Germany's Green Industrial Policies: The Costs and Benefits of Promoting Solar PV and Wind Energy,* Research Report, Geneva: International Institute for Sustainable Development.

McDowall, W., Ekins, P., Radosevic, S., and Zhang, L.-Y. (2013). 'The Development of Wind Power in China, Europe and the USA: How have Policies and Innovation System Activities Co-evolved?', *Technology Analysis and Strategic Management,* 25(2): 163–85.

MAKE (2013). *Europe, Middle East and Africa (EMEA),* Regional Report, MAKE Consult, Aug. Cited in 'De næste store vind-markeder boomer' (The next big wind markets are booming), Børsen, 20 Aug. 2013.

Mathews, J. (2012). 'Green China – Part I: Why Go Green?', *The Globalist,* 22 Nov.

Mathews, T., and Tan, H. (2014). 'China's Continuing Renewable Energy Revolution: Global Implications', *Asia-Pacific Journal,* 24 Mar.

Morrison, G., and Yoshida, P. (2009). 'China, United States, Korea Take Lead in Clean Energy and Low-Carbon Initiatives', *Research Technology Management,* 52(6): 2–4.

NAW (2014). 'Top 15 Wind Turbine Suppliers of 2013 Revealed', *North American Windpower,* <www.nawindpower.com> (20140311).

Qiu, Y. M., and Anadon, L. D. (2012). 'The Price of Wind Power in China during its Expansion: Technology Adoption, Learning-by-Doing, Economies of Scale, and Manufacturing Localization', *Energy Economics,* 34(3): 772–85.

REN21 (2012). *Renewables 2012: Global Status Report.* Paris: Renewable Energy Policy Network for the 21st Century, <http://ren21.net/Portals/0/documents/activities/gsr/GSR2012_low%20res_FINAL.pdf>.

Ru, P., Zhi, Q., Zhang, F., Zhong, X. T., Li, J. Q., and Su, J. (2012). 'Behind the Development of Technology: The Transition of Innovation Modes in China's Wind Turbine Manufacturing Industry', *Energy Policy,* 43: 58–69.

Schmitz, H. (2006). 'Learning and Earning in Global Garment and Footwear Chains', *European Journal of Development Research,* 18(4): 546–71.

Schmitz, H. (2015). 'Green Transformation: Is there a Fast Track?', in I. Scoones, M. Leach, and P. Newell (eds), *The Politics of Green Transformations,* London: Earthscan Routledge.

Schmitz, H., and Strambach, S. (2009). 'The Organisational Decomposition of Innovation and Global Distribution of Innovative Activities: Insights and Research Agenda', *International Journal of Technological Learning, Innovation and Development,* 2(4): 231–49.

Schmitz, H., Johnson, O., and Altenburg, T. (2015). 'Rent Management: The Heart of Green Industrial Policy', *New Political Economy,* forthcoming.

Sørensen, B. H., and Springborg, S. (2010). 'Sidste udkald for dansk vindindustri' (Last call for the Danish Wind Industry), *Berlingske Business,* <www.business.dk/energi-miljoe/sidste-udkald-dansk-vindindustri>.

Sturgeon, T. (2002). 'Modular Production Networks: A New American Model of Industrial Organization', *Industrial and Corporate Change,* 11(3): 451–96.

Venables, A. J., and Yueh, L. (2006). *The China Effect,* Centre Piece 208, Centre for Economic Performance, London School of Economics and Political Science, <http://cep.lse.ac.uk/pubs/download/CP208.pdf>.

Wang, Z. Y., Qin, H. Y., and Lewis, J. I. (2012). 'China's Wind Power Industry: Policy Support, Technological Achievements, and Emerging Challenges', *Energy Policy,* 51: 80–8.

Watson, J., and Wang, T. (2007). *Who Owns China's Carbon Emissions?,* Tyndall Centre Briefing Note 23, Norwich: Tyndall Centre.

Wieczorek, A. J., Negro, S. O., Harmsen, R., Heimeriks, G. J., Luo, L., and Hekkert, M. P. (2013). 'A Review of the European Offshore Wind Innovation System', *Renewable and Sustainable Energy Reviews,* 26: 294–306.

WWEA (2014). 'Statistics of Wind Power Development in China 2013', *WWEA Quarterly Bulletin,* 1 Mar.: 22–33.

Zeller, T., and Bradsher, K. (2010). 'Wind Power for Boston, Made in China', *New York Times,* 16 Dec.: B4.

Zhou, Y. C., Zhang, B., Zou, J., Bi, J., and Wang, K. (2012). 'Joint R&D in Low-Carbon Technology Development in China: A Case Study of the Wind-Turbine Manufacturing Industry', *Energy Policy,* 46: 100–8.

6

Transition Paths

Assessing Conditions and Alternatives

Staffan Laestadius

6.1. Introduction

Persistently high unemployment is arguably the most serious indicator of the European crisis.[1] For several decades the unemployment rate has exceeded 7 per cent and in the autumn of 2014 it stood at over 10 per cent in Europe as a whole and 11 per cent in the euro-area. Two countries—Spain and Greece—have over double the European average. Youth unemployment is typically twice the average unemployment rate in most countries. Europe is continuously accumulating young people with no, or at best marginal, connections to the labour market—a potentially explosive political and social situation in the longer term.

The accumulation of greenhouse gases (GHG) in the atmosphere is much less visible. The consequences on climate and for the planet are more insidious and less obvious, even in periods of high temperature, stormy weather, or floods. All European countries, however, currently contribute to global CO_2 emissions substantially more than is sustainable in the long run, as well as more than the global average, at least if calculated on a per capita basis (IEA 2013a, 2014a, c).

To have a reasonable chance (50 per cent) of limiting the global temperature increase to around 2°C during this century, we have to reduce global GHG emissions by at least 30 per cent before 2035.[2] Indeed, many analysts argue that we have to reduce emissions even faster and to a greater extent—say by

[1] This chapter is based on a book in Swedish, the translated title of which is *Climate and Welfare: Towards a New Swedish Model* (Laestadius, 2013a).

[2] This conclusion is based on the following sources: IEA, 2013a, 2014a, d; IPCC, 2013, 2014a, b; UNEP, 2014; World Bank, 2012).

50 per cent—if we are to have a real chance of meeting the 2°C target, since the widely discussed limit of 450 ppm CO_2 equivalents in the atmosphere is far from safe.[3] And we are already dangerously close to that level (Rockström et al., 2009; Hansen et al., 2008, 2012). Even a 1°C increase—which it is now probably too late to escape—may have a serious impact on many areas of the Earth (Hansen et al., 2012; IPCC, 2013).

For a long time there has been a strong link between economic growth and the growth of energy consumption in general and the use of fossil fuels in particular. Since around 1820 when the steam engine first became widely used (von Tunzelmann, 1978; Berg, 1985), the industrial metabolism—and thus our employment and lifestyles—has become addicted to the burning of more and more fossil fuel (Brondel, 1976; Landes, 1969; Maddison, 2007; Stern, 2007). This now corresponds to 82 per cent of global energy production and approximately 70 per cent of GHG emissions (IEA, 2014b).

The immediate consequence of this persistent connection between energy and growth is that in a world of business as usual (BAU) there is a direct conflict between climate change mitigation, on the one hand, and the elimination of unemployment—and our modern way of life—on the other. Unless we can 'pump up demand' in Europe in a sustainable way—as discussed, among others, by Stern (2008)—we will become stuck in a 'Scylla and

Box 6.1. OUR ADDICTION TO CARBON

Since the early nineteenth century, a 3 per cent annual global increase in carbon-based energy production has been perhaps the main lubricant for achieving freedom and modernity in advanced industrial countries. Travelling between academic conferences, escaping from Northern winters, migrating to creative cities, and performing the middle-class-dominated morning and afternoon shuttling between schools, kindergarten, violin classes, sports activities, and divorced parents—all this is based on burning carbon, as indeed is our access to cheap smartphones, with the CO_2 emissions from global mobile networks and ICT systems now representing a similar magnitude of CO_2 emissions to global air transport or cement production (Fehske et al., 2011; Malmodin, 2010).

This penetration of carbon into most aspects of our lives conditions our understanding of climate change. Arguably, as Giddens (2009) and Norgaard (2011), for example, point out, awareness of climate change is not so much a question of information: people already know (or know how to acquire the relevant knowledge), but they find severe difficulty in handling the discrepancy between their modern lifestyles and somewhat remote threats in the future. The challenge is therefore about gaining acceptance and legitimacy for a transformation policy we are all rather reluctant to hear about.

[3] The concept CO_2 equivalent (CO_2e) is a means to transform the climate impact of all GHGs into a common yardstick. As long as the main focus in the analysis is on CO_2, which is the case in this chapter, this distinction is of less relevance.

Charybdis' situation, in which moving away from one crisis takes us closer to another. In addition, climate change is a truly global phenomenon and hence only partially amenable to European policy and action.

Assuming that other countries will follow, Europe should reduce its CO_2 emissions to roughly half the present level within two decades to avert a temperature rise of more than 2°C. Not least in countries like Croatia, Cyprus, Greece, Italy, Portugal, Slovakia, and Spain, it will be difficult to mobilize political support for climate change abatement when a third or even in some cases more than half the young generation is unemployed. There will also in many countries be a fear of what climate mitigation may lead to among those already employed, unless there are policies that convincingly address both problems—unemployment *and* the climate. This is, for example, the case with Poland with large employment in the coal-mining and energy sectors.

This tough message can be translated into the following: after two centuries of carbon-based energy consumption growing at 3 per cent annually (Maddison, 2007; Smil, 2010), we must instead learn to reduce our addiction to carbon by approximately 3–4 per cent a year for the decades ahead, coupled with an increase of employment-creating activities of a similar magnitude. No country in the world has any experience of a structural transformation of that kind and magnitude, especially as a significant part of our productivity increase historically has been the result of human efforts—*brain power*—to utilize a growing amount of *external power*—primarily in the form of fossil fuels, which over a long period have become ever cheaper (Brondel, 1976; Landes, 1969/77; van der Wee, 1987). As was once observed to be the case for *technology* (Abramowitz, 1956; Solow, 1957), the importance of *energy* inputs in the production function has for too long been neglected.

This chapter assesses the potential for transforming the European economy away from its present unsustainable fossil-fuel trajectory in conjunction with a reduction in unemployment levels. Although the focus is on Europe, the issue is of global relevance. The chapter is structured as follows: a short section 6.2 attempts to identify the conditions for such a transformation. Section 6.3 contains an illustration of what an industrial transformation towards sustainability might entail. The role of innovation and new technology is analysed in section 6.4, followed in section 6.5 by an analysis on the global character of the problems we face. The policy options for crisis solution are analysed in section 6.6, while section 6.7 summarizes the main conclusions.

6.2. The Conditions for Transformation

While we may know from where we have to move, we do not know exactly where we need to go. Consequently, like Alice in Wonderland, we do not know what road—or roads—to choose. That is not necessarily a problem: the structural transformations undergone since the Industrial Revolution were not controlled in their details and goals, but occurred with a certain amount of self-organization, although within an evolving institutional framework favourable for entrepreneurship and innovation (Landes, 1969). What we are talking about now is, likewise, finding an institutional framework that, on the one hand, imposes strong restrictions on abuse of the climate and, on the other, creates powerful incentives for industrial creativity and transformation away from the carbon-based economy. In other words, Prometheus must be partly fettered once more, in parallel with providing incentives for the titan to contribute to transforming the economy based on creative destruction of 'black' (i.e. GHG-emitting) sectors and technologies. Basically this is an argument for more state intervention and guidance in the capitalist economy, as argued in Chapter 9 by Mazzucato and Perez.

Following publication of the Stern (2007) report, a debate arose among economists on what discount rate to choose when analysing the time dimension and character of climate mitigation and adaption paths (e.g. Gardiner, 2011; Nordhaus, 2007; Pindyck, 2012). The question raised is whether it would be less costly to postpone these measures for another decade or two (Tol, 2009). However, to have a realistic chance of negotiating the transformation

Box 6.2. THE TRANSFORMATION CHALLENGE

The conditions for transformation can be formulated in macroeconomic terms using a three-sector model:

* *green sector* (sustainable)
* *blue sector* (sustainability enabling)
* *black sector* (climate impacting sector).

The *green sector* consists of non-CO_2 generating activities (the service sector, ecologically produced food, and other 'green artefacts'), investments in human and cultural capital, and exports of non-CO_2-emitting goods and services. Bio-based products from sustainable forestry and agriculture also belong to this category.

The *blue sector* consists of activities which in themselves may be polluting but which in their use will contribute to a reduction of CO_2 emissions. On the consumption side this includes energy-saving and long-lasting artefacts (e.g. heat pumps). Among blue investments we may identify railway tracks as well as wind power systems and investments in energy efficiency.

To the *black sector* we allocate all consumption of—and investment in—products, systems, and services that increase CO_2 emissions.

The policy challenge from a macroeconomic point of view is not only to 'pump up demand' to increase employment but also, with help from the *blue* sector, to rapidly transform demand away from the *black* sector towards the *green*. It is not enough to reduce the relative size of the *black* sector; it must be reduced in absolute terms, eventually to a negligible level.

ahead of us, we must, for technological, economic, and cultural reasons, start immediately. The transformation—even if extremely rapid—will take decades to work through, and the process has to gain legitimacy and acceptance among people all around the world. Significant results must be achieved before 2030, and even before 2020 if we are to have a chance of not overshooting the emission targets. As it is the accumulated GHGs that count and the present emissions are much higher than they were two decades ago, speeding up the transformation process will significantly reduce the long-term burden.[4]

* The *technological dimension* can be illustrated by the finite character of the capacity of our Earth to absorb CO_2. To have a 50 per cent chance of keeping the global temperature increase below the 2° target there is a probable limit of further CO_2 emissions in the magnitude of 1100 Gt before 2050 (IEA, 2013a). To make the transformation manageable—in the magnitude of 3–4 per cent/year starting from present emission levels—it is necessary to achieve significant results before 2030 of mitigation activities combining transformation away from fossil fuels and energy efficiency measures (see also IEA, 2013b; IPCC, 2014a, ch. 6).
* The *economic dimension* can be illustrated with the economic lifetimes of the fossil-burning capital structure owned by firms as well as private individuals. Once installed or acquired, these are sunk costs providing cheap services over a long period—invested capital that owners will fight to defend from obsolescence. Those who buy an SUV today will not accept that their large investment is not to be used tomorrow. And once the car is purchased, it will be difficult to rapidly price its use out of the market. These processes must be allowed to take time; which will be all the shorter if we postpone the start of the process.
* The *cultural dimension* relates to the modernity problem briefly mentioned in Box 6.1 above. Transforming our lifestyles, which will be necessary at least to some extent, and redefining our views on the fundamentals of welfare, is not something that can be implemented instantly, for example, in autumn 2030. Values change over decades—some historians would even say centuries—and how to accelerate that process will probably be the most difficult part of policy, not fixing the technology.

The fundamental time perspective, then, is less related to how to construct an optimal transformation path but more how to achieve, and gain acceptance for, a fast enough take-off. Although the transformation will differ between countries and regions, due to their location as well as their economic,

[4] For references, see n. 2.

industrial, and social circumstances, a few sectoral aspects of a more general nature may be identified:

* As clearly stated in IEA reports (e.g. IEA, 2013a), the transformation will not be possible without a significant reduction in activity levels of energy-consuming systems, on the one hand, and improvements in the energy efficiency of those systems, on the other. 'Technology fix' solutions involving the substitution of solar, wind, and biomass for coal and oil, although necessary, will not be sufficient, given the present level of energy 'consumption'.[5] A rapid increase in sustainable forms of energy must be accompanied by a significant reduction of energy demand.
* This transformation of energy use relates not only to fossil-based systems. Although fossil fuels presently constitute 82 per cent of global energy supplies, the energy system is basically an open system, in which one source of energy can substitute directly or indirectly for another. In short: to facilitate the transformation, energy-using systems based on renewable energy also have to economize their energy needs.
* This global transformation is likely to reinforce the global scramble for biomass. As biomass—when sustainably produced and harvested—is an attractive resource for many activities, there will be fierce competition between using it for eating (and feeding livestock), drinking, driving, construction, packaging, bio-refineries and various other purposes (Novotny & Laestadius, 2014). The annual global production of biomass will be far from sufficient to satisfy all that demand (Canadell & Schulze, 2014; Running, 2012; Smith et al., 2014). In Europe alone, the forecast demand for biomass will in the immediate future be approximately 50 per cent more than its annual growth, hampering growth in the biomass-based sectors as well as increasing the pressure on non-European forests (UNECE/FAO, 2009, 2011). So, even if European forests are growing, forests all round the world will probably need more protection against deforestation than ever before.
* The mirror image of the transformation away from carbon addiction and the black sectors of the economy is a sharp reduction in the material obesity of the global economy. The world economy has, in fact, never been as 'heavy'—measured in tons of material handled in the economic metabolism—as it is today. The magnitude and rate of increase of iron, copper, oil, coal, aluminium, and cement production is not only higher than ever; the annual increase of several of these bulk materials is greater

[5] Energy is strictly not 'consumed' in its use by mankind but transformed to other—generally less useful—forms (although each transformation is accompanied by a certain 'loss' of useable energy).

today than was their total use only a few decades ago (USGS, 2014). Much of this has its origin or destination in China. That does not change the overall picture, however: on the one hand, China is now doing what other parts of the world have done before; on the other, Chinese industries now transform materials, much of which have their final destination in the old industrial world.

* Here is also perhaps the biggest contradiction in the present global situation: all policies aiming to stimulate demand in order to increase economic activity and reduce unemployment will, *ceteris paribus*, worsen the climate situation. Although there may be cases when economic growth can generate resources which, in a second round, will be useful in solving pollution problems, this is in general not the situation as regards climate change. Basically, economic growth, with its present demand structure, makes things worse, even if we can see that the relative energy or CO_2 content per € GDP is declining. All that count are the absolute emission levels.[6] And emission-free growth is what we, at best, can obtain in the green sectors, sometimes helped by the blue ones.

The transformation to green and blue activities may also cause a productivity dilemma. The increase of total factor productivity originates, as mentioned in section 6.1, in the combination of the inputs of *brain power* and *external power* (energy). The first may be summarized as follows: based on our creativity, humans can produce technical as well as organizational innovations, making it possible to perform the same work, or work of higher quality, with less use of energy; or we can increase production using less energy. The wheel was such an innovation, reducing friction and making it possible to increase the amount pulled by a horse. A thin-bladed saw of high-quality steel reduces the energy needed as well as the material loss when cutting up a timber log. An electric engine typically has twice the energy efficiency of a combustion engine. These productivity increases illustrate the direct use of knowledge and creativity.

To this we have to add that technological development, based on the substitution of energy for labour, *ceteris paribus*, necessitates *external energy*. Over time, capital-intensive industrial processes develop as brain power is used to construct energy-transforming process equipment that substitutes for labour, thereby making higher levels of energy-consuming material production possible. Here is the problem: a significant part of productivity increases has, since the origin of the Industrial Revolution, been based on increasing energy inputs (Landes, 1969). A large part of the increased use of fossil fuels—approximately 3 per cent annually for two centuries—has been used for productivity purposes. What will happen to future productivity—and

[6] In fact what matters are the *accumulated* GHG emissions.

thus economic growth—now that industrial transformation aims to move away from carbon?

Once again, we are back to the problem of whether climate change abatement may be looked upon as a headwind dramatically reducing the probability of economic growth (Gordon, 2012) or whether the transformation policies we are ultimately striving for will create the conditions for sustainable growth (Arrow et al., 1995).

6.3. Looking into the Details: This is What Transformation is about

A transformation policy aiming to provide employment opportunities and reduction of GHG emissions has to focus on a few development blocs. The *development bloc* concept, as formulated by the Swedish economist Eric Dahmén, relates to the dynamics of sequences of imbalances occurring when companies or innovators react to *necessities* and *opportunities* created in industries and technologies that are interdependent (Dahmén, 1950, 1989). In his Schumpeter-inspired 'causal analysis', Dahmén's explicit focus is on the transformation of industries and technologies—not on aggregate growth—and on the proactive or reactive responses made in the various blocs. Development bloc borders do not necessarily follow industry classifications nor simple value chain schemes; they may also shift over time as advances in technology change the business logic of industries.

Although most industrial or economic activities are more or less integrated in GHG-emitting systems, some are more heavily based on fossil fuels than others. Simplifying somewhat, we can identify two related development blocs that are crucial to the European reduction of CO_2 emissions as well as to employment creation: transport and energy. Large-scale transformation of these blocs may have significant multiplier and secondary effects on the rest of the European economy.

Within the *transport bloc*, there are certain low-hanging fruit which can be captured more or less immediately. Approximately 84 per cent of all personal transport in Europe (EU27) involves cars (Eurostat, 2014). Although it is far from obvious what the future transport system will look like and what mix will emerge from various technical solutions, it is obvious that car transport must be significantly reduced, and rapidly. A 25 per cent reduction of car transport within two decades—i.e. a reduction from approximately 84 per cent to 63 per cent, given constant transport volume—will require a doubling of public transport all over Europe. That will create new employment, new business solutions, large investments, and transformation of the automotive sector to new forms of vehicles; buses and trains will increase in importance. This may provide a dynamic spirit to European industry, enhancing its competitiveness.

Although there are variations among European countries, the short-term transformation of personal transport should include many more buses and coaches: the roads are there already and the impact of investment in buses is immediate, whether on CO_2 emissions, on the transport system, or on employment. Data from Sweden reveal that investment in coaches could double immediately and remain at that level for decades (Laestadius, 2013a).

The *energy* bloc is more complex, although the relative magnitude of the challenge is similar: most European countries have an energy mix where fossil fuels make up more than 80 per cent. The present low prices of fossil fuels—due to their non-priced externalities as well as the recent growth of fracking—in combination with the instability of renewable forms of energy creates major challenges for policy-makers. To gain acceptance and create incentives for energy transformation, there is a need for long-term changes in taxes, subsidies, and prices, as discussed in Chapter 7 by Lauber and Jacobsson. New forms of energy management will also need to be developed to handle imbalances in the future energy system.

Balancing the energy system is not a fundamental problem, however, unless we make it so (IEA, 2014b). As shown in Box 6.3, the potential for renewables in Europe is enormous, and most countries can—if there is sufficient political will—induce wind or solar power which, via feed-in systems or certificates, can reduce dependence on carbon by another 20 per cent. At the margin, this will increase energy prices somewhat: this is exactly what is needed to induce large-scale energy efficiency programmes all over Europe. On average, 20 per cent higher energy prices can be compensated for by the low-hanging fruit in energy saving—an activity which will largely be performed in another development bloc: *building and construction*. In addition, this energy saving will create employment.

The transformation of the energy system towards sustainability is related to an enormous investment programme in electricity transmission and in new technologies for electricity management and control. European countries have to be more interconnected than hitherto and trade more in energy. This will reduce CO_2 emissions and create new businesses as well as reducing the vulnerability of the European energy system. These investments will also be employment creating.

6.4. The Role of Innovation and New Technology

The conventional wisdom—particularly among policy-makers—is that climate change has to be mitigated primarily with the help of science and technology

Box 6.3. BENCHMARKING THE NEAR-TIME TRANSFORMATION POTENTIAL: THE CASE OF WIND POWER

Although the transformation challenge is enormous in the energy bloc, there are certain low-hanging fruit to start with. Below, eleven European countries are benchmarked against Germany as regards installed wind power capacity, thus revealing the 'instalment gap' to be overcome in catching up with the present German wind power density. And that density is not necessarily the end-point. Germany is still adding new wind power capacity to its grid.

Country	*Fossils/Total energy cons.(%)*	*Wind power MW*	*Wind power MW/000 km²*	*Instalm. gap MW (rel. to German)*
Germany	82.9	33730	94.5	0
Austria	70.0	1684	20.1	7917 (0.21)
Denmark	80.0	4772	110.7	–
Czech Rep.	79.0	269	3.4	7477 (0.036)
Finland	56.1	448	1.3	32566 (0.014)
Greece	92.7	1865	14.1	12499 (0.15)
Hungary	81.3	329	3.5	8883 (0.037)
Italy	87.5	8551	28.4	28453 (0.30)
Poland	96.3	3390	10.8	29662 (0.11)
Spain	76.9	22959	45.4	47789 (0.48)
Sweden	30.8	4470	9.9	42668 (0.10)
UK	87.5	10531	43.0	23144 (0.45)

Sources: BP, 2014; Eurostat, 2014; EWEA, 2014.

(S&T). In certain policy circles there are strong expectations of some form of 'technology fix' and of science-based solutions which, it is assumed, will alleviate the problems of the enormous transformation we now face. This faith in science and technology is firmly rooted in our history of enlightenment and modernity (Maddison, 2007; Mokyr, 2009; Toulmin, 1990).

The widespread acceptance of the long-term importance of science sometimes leads to an exaggeration of the role of science in individual industries and technologies as well as to oversimplified models of how enlightened ways of thinking penetrate industry and technology. The well-known 'linear model'—once critically scrutinized by Kline & Rosenberg (1986)—illustrates that phenomenon. The linear perspective has for a long time influenced technology policy within the EU as well as the broader OECD area. If science a decade ago was part of European policy for competitiveness (the Lisbon strategy), it is now also looked upon by many as the main tool for climate policy. Undoubtedly more S&T activities must be stimulated in grappling with the challenges of the industrial transformation ahead. We should not, however, underestimate the achievements of S&T in earlier decades or centuries, nor the interactive learning processes that may take place once we commence large-scale investment in technologies we already basically know (see Chapter 7 by Lauber and Jacobsson in this volume). In short, we know a lot already to help us transform.

Innovations, following Schumpeter, are new and creative combinations of recently developed as well as existing knowledge. The low-hanging fruit of innovative solutions in climate mitigation adaption, for decades ahead, have to be found mainly in what we already know and in the best practices of today. Significant creative recombinations are needed; but not just those based on recent technological breakthroughs. Such breakthroughs—like carbon capture and storage (CCS), fourth-generation nuclear power, and advanced batteries—will probably be too late or too expensive to place much faith in during the transformation ahead.[7]

In addition, most CO_2 emissions outside the transport sector have their origin in low- and medium-tech sectors and technologies. These are often characterized by high material volume and low R&D intensity. Scientific breakthroughs are unlikely to reduce CO_2 emissions from the cement industry, for instance—significant reduction of its use in construction, along with mixing cement with strength-enhancing components will. The situation is similar for steel. Fundamental scientific breakthroughs within these sectors or technologies are probably less important than a reduction in the use of steel, increased recycling, and growing use of renewably produced electricity in steel-making (Laestadius, 2013b).

Prices are arguably important here. Many renewable technologies will still not be competitive as long as fossil-based technologies do not pay for their climate impact. Innovation policies, therefore, have to include changing relative prices of energy. This requires significantly higher real prices for CO_2 emissions, challenging present low-priced fossil fuels in parts of Europe, but also supporting policies for renewables. Moreover, breaking away from the locked-in paths of modern lifestyles and social organization as discussed in section 6.2 may necessitate social, organizational, and institutional innovations—and not just technological ones.

6.5. The Global Dimension

The crises discussed here are genuinely global. Our dilemma is how to govern a global 'commons' sometimes described as 'slippery' (Markusen, 1996) or as a 'flat world' (Friedman, 2005). Whatever is done in one part of the world may

[7] A detailed description and analysis of CCS is to be found later in this chapter (Box 6.5).

have leakage effects in another. Although climate change may be regarded as the ultimate globalization phenomenon, it has, until recently, been largely neglected in the globalization discourse (e.g. Castells, 2000). In what follows, we focus on the global aspects of climate mitigation.

Global warming is a genuinely global phenomenon; although there may be time lags and complex global climate dynamics, what we release into the atmosphere will impact on all of us in one way or another. At its starkest, the apparent implication is that if (many) other countries and actors in those countries choose to ignore climate change, then there is not much one can do about it. The situation is, however, not necessarily that bad. We are facing an interactive game where actors all over the world (companies, NGOs, governments, IGOs) are positioning themselves in relation to climate change and are searching for reasons and possibilities to mobilize for action. Some are doing so for ideological reasons; others because they are worried about the climate; some because they are stakeholders in the transformation to sustainability.

Policy-makers engaged in creating conditions for the mitigation of climate change thus have to navigate through this. It is not primarily a question of knowledge: most governments know, or know how to get the knowledge, even if they do not always reveal this in their policies. US government bodies, for example, are probably the best sources for climate data and climate-related research and analysis.[8] Although there may be exceptions, many global corporations know what is going on. So while, on the one hand, knowledge is there, on the other, the necessary activities are often lacking.

Policy-makers fear that global leakage effects may make mitigation and abatement efforts appear like a drop in the ocean. Related to this, those who shoulder the burden of mitigation may have to share the benefits with everyone else on the planet. The demand for concerted global action is thus strong. The probability of achieving global, common, and far-reaching commitments is, however, very low. A globally coordinated climate policy revolution is unlikely in the immediate future.

Change must therefore have another point of departure; someone—some actor or reasonably coherent group of actors—must take the initiative. It is (almost) always possible to do at least something that changes part of the conditions for other actors in the system. Indeed, that is what we label as innovation: introducing new combinations, thus changing the rules of the game in one way or another. The 'others'—the initial 'non-innovators'—then

[8] Among the best known are the Carbon Dioxide Information Center (CDIAC, <http://cdiac.ornl.gov>) related to the US Department of Energy (DOE), the National Aeronautics and Space Administration (NASA) and its Goddard Institute for Space Studies (GISS), the National Oceanic and Atmospheric Administration (NOAA, <www.noaa.gov>), and the National Snow and Ice Data Center (NSIDC, <http://nsidc.org>).

have to react to the new conditions, contributing to a chain reaction of a more or less self-reinforcing character. This is what path-creation and path-dependence are all about (Garud & Karnoe, 2001).

Returning to climate policy and somewhat simplifying, we can identify two outcomes of national or regional climate initiatives: a 'leakage effect' and a 'snowball effect'. The former has been pointed out in many international analyses of climate change politics and contributes to the sceptical position—not the least among CO_2-emitting companies operating in global markets—against policy measures that deviate too far from global norms (e.g. Hoel, 2012). A fear in Sweden, for example, is that overly tough CO_2 conditions for Swedish steel producers may result in steel production moving to India or China where the relative CO_2 emissions per ton of steel produced are much higher. At the margin, global CO_2 emissions may therefore increase.

Hence, in a totally frictionless world, leakage effects may be significant. Any cost-driven policy will tend to move industrial activities to other parts of the world (Friedman, 2005). But the world is not totally frictionless nor totally amoral. Nor are all climate-mitigating initiatives unprofitable, not even in the short run and under current conditions (McKinsey, 2008b). Activities that imply major investment in physical and human capital will not be undertaken in the short term just because climate laws in other countries have not yet caught up. In addition, countries that tax negative climate impact effects can compensate the losers in various ways. The risks of short-sighted 'leakage effects' should not, therefore, be over-emphasized: ultimately firms have many other parameters to consider when formulating their strategies.

As for the 'snowball effect', the basic character of this effect is obvious: many actors realize that they have to engage in climate change mitigation or adaption activities. There is a space or niche, the size of which we cannot know in advance, available for innovative experiments searching for new paths of development. The rationale for this may be sought in demonstration effects, in game theory, and in the management of 'the commons' (Ostrom, 1990; Ostrom et al., 2002). This was also analysed by Stern (2007, part VI) and in a recent Swedish/Norwegian study (Hoel, 2012). The theory of the game—one might call it 'the climate walk'—is simple: I take a small step in a friendly direction and promise that I will take a second, provided you follow after me on the first. I take an initial risk and may even promise to take more risks in subsequent steps if you join the risk-taking walk. At best, that creates a dynamics of its own and may encourage third- and fourth-tier followers, relieved that someone else has taken the lead.

This is the essence of the 'snowball effect' argument. There is another related argument, the so-called 'Porter-van der Linde effect' (1995). This suggests there is a first-mover competitive advantage for countries that, due to tough environmental policies, force domestic firms to develop green technologies which, in a second period, can become profitable when other countries follow. However, the main objective for the policies discussed here is not so much to achieve local competitive advantage as to contribute to the transformation through a new and sustainable industrial and technological regime. The promises of the Porter-van der Linde argument may thus serve as a 'lubricant' in the process for achieving that goal.

6.6. Policy Options for a Crisis Solution

Let us summarize the conditions for policy-making so far. If we want to promote a global structural transformation away from a society with high unemployment and dependent on GHG emissions, it is necessary to move towards a robust international structure of prices, rules, and norms—a new institutional order—favouring such a structural change. International declarations not based on strong commitments will not be enough.

The effort within the UN system[9] to reach a global agreement involving all, or almost all, countries has hitherto not proved very successful and is unlikely to be so in the immediate future. Waiting for the last hesitant actor to join the transformation path away from CO_2 emissions is far too slow a process. Instead, policies must be initiated by countries and actors willing to launch a self-reinforcing process. This is a strategy based on the assumption, impossible to confirm in advance, that the positive dynamics of the stepwise snowball effect will be far more important than the leakage effects which will certainly also arise. In this section we analyse how such a policy can be formulated.

Even if seeking global agreement represents an unrealistic path on which to continue, we still have the option to start snowball policies at a *national level* as well as creating international alliances of *like-minded countries or other actors*: alliances that are small and similar enough to agree on the shape and magnitude of an initial snowball to start with but large enough to have an impact on the global community. We will return later to the international initiative, as our fifth building block. Instead, let us first focus on the national level, which also has to create firm foundations for any international snowball effect.

National policies can be formulated independently by nations, for example, by European governments in interaction with their citizens, and can also be

[9] This work takes place primarily within the United Nations Framework Convention on Climate Change (UNFCCC).

influenced by similar processes in other countries. These policies may be summarized in terms of four building blocks:

* an overall *climate policy commitment*, CPC;
* a far-reaching *climate taxation path*, CTP;
* a *climate innovation and investment policy*, CIIP;
* a *climate policy correction package*, CPCP.

Moreover, as mentioned, based on those packages we can also discuss a fifth element:

* an *International Climate Investment Community*, ICIC.

The first block—*a climate policy commitment* (CPC)—is probably the most strategically important. It may contain a far-reaching agreement to work for a transformation away from addiction to carbon in parallel with a restoration of full employment, a kind of Magna Carta where national policy-makers show that they have reached broad agreement on the nature and magnitude of the problems, although they are open to variations in the policy details. This may be a first step towards gaining legitimacy for far-reaching policies.

Part of the strategy in this climate policy commitment is to focus on *the rhetoric of crisis policy*. The transformation ahead will be far-reaching and long-lasting. Although it is important to start immediately with small concrete steps, the far-reaching consequences will only come after the low-hanging fruit have been harvested. And those long-term consequences must be discussed explicitly. Politicians must show they are planning for a transformation that will produce a welfare outcome quite different from, and intentionally better than, what 'business as usual' offers. This will be the tough part: a commitment to enforce the transformation and take care of the inevitable problems. In summary: the rhetoric must be based on policies to develop a Pareto optimal outcome from the transformation, convincing people they can be better off if they actively participate in that process. Basing policies on catastrophe scenarios and fear is probably not fruitful (Giddens, 2009; Urry, 2011).

This means that the overall paths for transformation should be agreed upon, and politicians should explicitly declare that fossil-fuel prices—indeed all energy prices—will increase little by little, that black sector activities will be gradually phased (and priced) out, and that this policy will include creating new employment in green and blue sectors while simultaneously reducing the black sectors. This will give people time to adjust and is also important in terms of creating domestic legitimacy for the international commitments discussed later.

If such a CPC relates to the *goal* of policy, the other three building blocks are concerned with the *means*. As regards the *climate taxation path* (CTP), we realize that policy-makers will differ as regards which path will be most effective to follow. That can hopefully be handled within the overall commitment. It is important, however, that basic agreement is reached on how to transform the tax and fee structure in each country to accelerate the transformation from carbon to sustainable forms of energy and sustainable industrial sectors. Although there may be variations between countries, one can imagine a toolbox of different taxes and fees to implement such a transformation policy. The detailed portfolio of taxes and fees can be fine-tuned once the process has started. The most important thing is probably to declare that the aim is to reduce black sector activities and that the thumbscrews will be tightened perhaps every half year. This will necessitate various forms of 'tax switching', i.e. compensating for carbon-focused tax increases with tax reductions in other parts of the economy. Strong leverage effects can be obtained with the aid of *bonus malus* (reward and penalization) systems. Value-added taxes can be differentiated between 'green' and 'black' consumption. And taxes on labour can be reduced to compensate for higher taxes on 'climate bads'. In short, taxation policy cannot stay neutral with respect to different forms of demand, but has to explicitly influence it, even though this runs counter to conventional wisdom in many finance ministries.

The *climate innovation and investment policy* (CIIP) is looked upon here as being highly integrated: the transformation suggested is largely based on significant investments in the blue sector, with innovative efforts being directed towards that sector. Although government R&D spending may be important in these areas, the fundamental incentives for innovations—R&D-based as well as others—must be found in government stimulated infrastructural investments, on the one hand, backed by new and emerging relative prices, on the other. Swedish experiences from the 1970s onwards reveal that, as long as prices (and expected prices) for non-sustainable solutions are too low, alternative technologies will never be introduced on a commercial scale, and R&D efforts will remain too low.

It may thus be argued that, unless CO_2 prices can be raised from recent low levels (throughout 2014 they were less than €7/ton) towards a sustainable level of perhaps €40/ton, there will be neither breakthroughs in CCS technologies nor any significant transformation towards sustainable forms of energy. New price relations are necessary to provide the innovative space for the creative destruction of non-sustainable technologies and systems.

Higher prices are far from sufficient, however. The price mechanism is not that efficient in the short run—say, two or three decades ahead—and may also create perverse incentives and windfall profits (Azar & Sandén, 2011; Bergek & Jacobsson, 2010). Complementary policy measures like regulations, administrative rules, and public purchasing will also be needed. The incandescent lamp was not phased out with prices but with regulation.

Box 6.4. WHAT IS THE 'RIGHT' PRICE FOR CO_2?

The price of CO_2 in the European Trading System (see Chapter 8 by Begg) did not exceed €7/ton during 2014. As a consequence of that and of low price forecasts, the transformation effect of ETS has probably been negligible. But what is the 'right' price for CO_2? Below, we discuss two ways to arrive at an estimate of what might be a reasonably 'right' price and price development path:

The externalities/social cost track: this line of reasoning with its origins in Marshall (1890) and Pigou (1920) involves identifying the negative and positive non-priced effects in society and estimating their magnitude. The approach faces three main problems:

1. If processes and consequences are allocated over time, the calculus has to include a model for discounting, which brings with it the problem of selecting a social interest rate (see e.g. Nordhaus, 2007; Pindyck, 2012; Stern, 2007).
2. There is always a problem of what impacts to include in the calculus. Is biodiversity and the 'beauty of nature' part of the calculus (Gardiner, 2011) or just the GDP level as in Tol (2009)? How should deaths and injuries that follow from CO_2 emissions be priced (Parry et al., 2014)?
3. How do we factor in the high risk of some fundamental threat to mankind or human habitats? In the climate change context, this is something that cannot be neglected.

Nevertheless, there is an established literature attempting to estimate the costs of carbon (e.g. Nordhaus, 2011, 2013; Parry et al., 2014; Stern, 2007; Tol, 2009). An overview of the literature is available in van den Bergh & Botzen (2014), who also conclude that all estimates below US$125/ton (€100) are unrealistic. There are, of course, many paths to increasing the CO_2 price. Nordhaus (2013), for example, argues for an immediate price level (2015) of US$25/ton followed by a 5 per cent yearly increase reaching US$53 (around €42) in 2030 and US$93 (or €74) in 2040. See also the estimates by Lauber and Jacobsson (Chapter 7) in this volume.

The transformation track: this line of reasoning starts with the assumption that we have to transform if we aim to stay below a certain target (normally the 2°C target). The argument (e.g. van den Bergh & Botzen, 2014) focuses on analysing price levels and paths that will provide price incentives for investors and other economic actors to eliminate CO_2 emissions. Two cost discussions are relevant here:

1. *The cost of renewable energy* that can substitute for carbon. Although there has recently been a significant cost decrease in wind power as well photovoltaics, there is still a cost gap between renewables and coal-fired plants. A CO_2 price of around €40/ton will, however, make coastal wind power cost-competitive with coal and natural gas-fired plants. For offshore wind power and photovoltaics, the CO_2 price needs to be higher. A typical coal-fired energy plant produces 900kg CO_2/MWh produced (IEA, 2010). A CO_2 price of €40–60 will increase coal-generated electricity prices by €0.036–0.054/KWh, making most forms of renewable energy competitive (see Kost et al., 2013, 2).
2. *The cost of CCS technology*. As shown in Box 6.5, there is a large range of cost estimates—typically from US$55–120 (or €44–100) t/$CO_2$ abated—depending on type of installation, although levels as high as US$250 t/$CO_2$ are also mentioned.

Part of the investment policy is to create better conditions for financing sustainable investments. Previously, finance markets have tended to underestimate the risks with fossil-based activities (see e.g. Carbon Tracker, 2014) as well as being sceptical towards and badly informed about green investment strategies. The investment policy must help to mobilize capital for investment in 'blue' infrastructure. This is an important role for the entrepreneurial state (see Chapter 9 by Mazzucato and Perez).

And so to the fourth element: the *Climate Policy Correction Package* (CPCP). From a technical perspective, this could be regarded as part of the other packages. But the technical character of the policy tools grouped under this umbrella heading is less important than their visibility and contribution to achieving trust, acceptance, and legitimacy for the transformation.

Structural transformations may well create losers and certainly have done so historically. Even if analysts find, ex post, that transformations are generally welfare-creating, the winners may be different from the losers. There is also evidence that recent economic development has not only increased unemployment and inequality but also brought about a decline in net incomes for large sections of the population in many countries.[10] Without digging too deep into welfare statistics, we may therefore assume that many in the European population may well feel threatened, say, by a policy not aiming to restore demand for cars and thus employment in the automotive industry but rather to reduce that industry by half.

What will be needed is a clear policy package explicitly addressing the challenges that will follow from the transformation policy. The list of possible temporary policies, rules, subsidies, and lubricating measures may well be long. Those who live in the wrong place or the wrong house, or those who have recently bought the wrong car or have a job or profession related to the black sector, should be given the opportunity and resources to escape from their lock-in situation and re-establish themselves in green sectors.

The broadly successful Swedish experiences in the decades after 1945 show that it is possible to carry out—and gain acceptance for—a radical policy of structural transformation where people move between industries, firms, professions, homes, and regions. The promise in the 1960s, for example, was that the effects of closing down low productivity firms and industries and moving into high productivity sectors would reduce unemployment and hence be beneficial for everyone. It might be argued that a Scandinavian model for

[10] The core issues of relevance here are, on the one hand, the dramatic development of IT, leading to a digital economy transforming labour markets (Brynjolfsson & McAfee, 2013; Stiglitz, 2014) and, on the other, the polarization of income and capital in advanced countries (Piketty, 2013; see also Ch. 9 in this volume). This parallel crisis contributes to a serious policy communication problem: abstaining from climate-related transformation will not be a solution to the challenges of the digital economy.

capitalist organization cannot be simply exported to other, larger countries (e.g. Erixon, 2003; Esping-Andersen, 1990; Hilson, 2008; Lundberg, 1985). Nevertheless, North European experiences (not least those of Sweden) reveal that there is the potential to win acceptance for radical policies, provided there is a reasonable institutional set-up connected to a welfare vision. Formulating such a green vision of post-carbon welfare represents an important European project.

The fifth policy package—the *International Climate Investment Community* (ICIC)—assumes that a small set of countries can agree upon a common 'snowball' to start with.[11] This international snowball must have its foundations firmly based in domestic commitments among participating nations and may be less far-reaching than some participants might desire. Most important, however, is that a small number of EU Member States, based on the new provisions of the Lisbon Treaty, agree a minimum platform for joint climate action rather than the exact magnitude of that action. This club or community may become an attractor, mobilizing second-mover initiatives by others. Growing consensus that climate mitigation efforts are essential and the recognition that many investments in sustainability are profitable should reduce the risks of free riders. It is important the ICIC is created in such a way that it cannot be blamed for protecting its own short-term interests. To engage members and support and further reduce the risks of free-riding, the ICIC may condition some of their policies so that they are valid only if they are not undermined by other countries. As long as there are important followers, the community can promise to continue the policy walk on the path away from carbon addiction.

A core part of the ICIC is an agreement to allocate significant resources for investment in renewable technologies and systems, to reduce unemployment in parallel with the reduction of CO_2 emissions. The important issue here is not the exact magnitude of the investments agreed upon—these may well vary across countries; what are important are joint public declarations that will force policy-makers to meet their promises and create expectations that have to be fulfilled. The ICIC should also take care of some of the leakage effects that may occur in parallel with creating various dynamic effects among participating countries.

Probably essential for the ICIC is a commitment from Member States to establish a *price floor* and a minimum *price development path* for CO_2 of approximately €40/ton CO_2. That should be a technology-neutral policy based on

[11] The idea of an International Climate Investment Community (ICIC) was originally proposed by the Swedish think-tank, Global Utmaning (Global Challenge), of which the author is a member. For the original proposal, see Larsson et al. (2010). What is proposed here is based on those ideas.

a commitment to impose CO_2 taxes or trading systems that are more far-reaching than the present ETS (see Chapter 8 by Begg), which for the immediate future will probably not be capable of increasing prices much above €15. That is just a third of what is probably needed to have a real effect on European transformation. In addition, member countries should agree to eliminate all direct and indirect subsidies for fossil fuels relatively quickly. Those commitments will create the institutional and economic conditions necessary to mobilize the sustainability-enabling (blue) investments needed.

Countries may deviate considerably in their detailed national policies. Indeed, it may be an advantage if there is such variation, indicating that there are several possible roads away from the present lock-in even within an international framework. The technical issues of the ICIC need not be analysed in detail here. Suffice it to say that, first, it is possible to combine the present ETS, which has more modest ambitions, with the more advanced system suggested in this chapter. Second, the level and format of the declared price development path for CO_2 is crucial. Irrespective of whether the ambition is to substitute sustainable forms of energy for carbon-based ones or to develop CCS technologies—which will be costly—it is necessary to embark on a path leading towards a price level for energy in general and fossil fuels in particular that is considerably higher than today. Although technological development may change conditions in the future, the overview in Box 6.4 indicates that a sustainable CO_2 price of around €40/ton is necessary to initiate substantial investment in new technologies for energy efficiency as well as for renewable energy systems. That is also roughly the price level needed to finance CCS (Finkenrath, 2011; MottMacdonald, 2012).[12]

That price level will also contribute to a dynamic transformation of the global energy system. All sustainable forms of energy will benefit if fossil-fuel prices are increased via CO_2 taxes and trading systems. Photovoltaics and offshore wind power—each with only marginal external effects—will become competitive on their own merits. CCS technology will, in its commercial form, be large scale, facing strong competition from sustainable forms of energy. It will be used primarily where CO_2 emission is most difficult to avoid, such as in the steel and cement industries. CCS systems will never directly solve the CO_2 emissions emerging from cars or other distributed sources of CO_2 emissions. Even if we continue to use fossil fuels in the future, the automotive system still has to transform itself by switching to vehicles powered mainly by renewable

[12] The low prices of fossil fuels which have dominated markets during 2014 and 2015 may delay this transformation. However, this is also an opportunity for governments/the EU to increase CO_2 taxes and prices when fossil fuels are cheap—and to maintain them on that level when oil prices increase again.

Box 6.5. CARBON CAPTURE AND STORAGE (CCS): A *DEUS EX MACHINA* AND THE ULTIMATE ENGINEERING PROJECT, OR JUST A PIPE-DREAM?

What is CCS?: the idea behind CCS is to capture the CO_2 produced in the combustion of fossil fuels, return it underground, and store it indefinitely in suitable geological sites until it is transformed into calcium carbonate.

The technology: the development of CCS technology follows three paths, none of which has yet become commercially available for power plants or CO_2-emitting industries. CCS involves the separation of CO_2 from industrial exhausts (a process which, in itself, consumes energy), transport to a suitable geological site, and 'permanent' storage deep underground (IPCC, 2005; McKinsey 2008a; Watson 2012).

The magnitude: globally there are approximately 9,000 large CO_2 emission sources, of which 7,000 are power plants and 2,000 steel works, refineries, and cement factories, all in need of CCS (Finkenrath, 2011; IPCC, 2005). Another way to formulate the challenge is as follows: the weight of the CO_2 molecule is approx 3.67 times that of the carbon atom or approximately three times that for fossil fuel. Considering the increase of energy which follows from the CCS system itself (over 25 per cent), this means that CCS logistics will involve more than 30 Gt/year, or over seven times the level of global oil and gas transported today (Laestadius, 2013a).

The costs: cost estimates, including those for transportation and storage, vary widely, depending on assumptions about learning processes. IEA estimates the costs to be US $55–80/ton CO_2-avoided (Finkenrath, 2011). Al-Juaied and Whitmore (2009) arrive at costs for CO_2-avoided in the range of US$35–150 excluding transport and storage costs. The industry itself has published estimates ranging from $23 to $106 (GCCSI, 2011). A detailed UK analysis, strongly stressing the uncertainties, ends up with total costs of around €100/ton of avoided CO_2 (MottMacdonald, 2012). A recent document, also from the 'CCS industry', indicates a range of US$70–250 t/$CO_2$ abated (= €56–200) depending on the type of installation. Installed in a new coal-fired generation plant, the estimate is US$90–120/ton (or €72–96) (ICO$_2$N, 2014).

Current status of CCS projects: although CCS has been developed and discussed for two decades, its present contribution—except for enhanced oil recovery (EOR, i.e. returning CO_2 back into the oil or natural gas well to increase the pressure)—is negligible. At the end of 2013, only four large-scale CCS systems intended for geological storage were in operation, while another nine were under construction (IEA, 2014a). This is far below what is needed if CCS is to play a significant role in the transformation away from carbon. If all today's planned CCS capacity is realized, it will still account for less than 1 per cent of power sector CO_2 emissions in 2012 (IEA, 2013a).

electricity. As mentioned above: there will simply not be enough biofuels available to put new fuels in all old tanks.

Finally, the ICIC will initially generate significant financial streams when collecting CO_2 taxes or selling new forms of certificates. To gain acceptance for the community idea, it is probably better to return those revenues to participating countries for national decisions on how best to use them to obtain the greatest leverage effects. At best, participating countries may agree to allocate those funds to blue sectors of the economy, generating employment through investments in sustainable infrastructure.

In its ideal form, international fuel bunkers should be included in an ICIC.[13] The recent problem of trying to win international acceptance for a pan-European tax on aviation fuel indicates the magnitude of the challenge of trying to gain acceptance among stakeholders. If a European bunker tax of €20/ton CO_2 is imposed, that will initially create annual revenues of around €5 billion, which then have to be distributed in some form. When raised to the more long-term sustainable level of €40/ton, the financial flows may not increase much more, particularly if European air transport is reduced to half its present level and/or shifts to biofuels.

The ICIC addresses the challenges in many dimensions. It indicates the direction of travel, creates a price floor for investment, and encourages significant structural change from the black towards the blue and green sectors of the economy. It also indicates the existence of a common platform for action among countries. In short, it may—if suitably reinforced by national policies—serve as the first roll-out of a small international 'snowball'. The aim is for it then to become an attractor for other countries to join in.

6.7. Conclusions: Is All This Possible?

It is far from obvious that a policy as outlined in this chapter will be successful. The rapid change in climate in recent years, in combination with the economic crises, not least in Europe and the USA, as well as the stalemate in international politics, reveals that the overall situation is very serious. While things may become worse before they start to get better, the conclusion, based on the research underpinning this chapter, is that there is still a narrow path open for crisis abatement. The path set out here is based on the assumption that rational actors in an open world, faced by severe problems, will search for solutions enabling them to avert a tragedy of the commons. There is therefore space for innovative activities by countries and by communities of countries. And if reasonably well organized—for example, within Europe—this may have a beneficial snowball effect on global crisis policies and management. It is difficult to see any fundamental alternative to such a strategy.

The content of such policies is simple. Instead of aggregate demand, we need to focus on a strong structural transformation. Future employment and welfare must be created in a large-scale transformation towards the green and

[13] Fuels for international maritime transport and aviation are, according to international agreements, neither taxed nor included in national transport or energy statistics but are reported separately. Globally, international bunkers represent 2.6% of energy use, 7.9% on average of oil consumption, and approximately 3.4% of CO_2 emissions. In total, European bunkers constitute approximately 89.5 Mt of oil consumption, causing 288 Mt of CO_2 emissions (IEA, 2014a, c; Eurostat, 2014).

blue sectors. The black sectors must be dramatically reduced, in absolute as well as relative terms.

It is probably not the case that this transformation, even if employment-creating, can be combined with rapid economic growth. As mentioned, 'economic growth' is not the target for the policy analysed above: employment, transformation, and new forms of sustainable welfare are. All this may end up in 'green growth' (OECD, 2011; see also Chapter 9), although not necessarily.

If anything, climate change has revitalized the classic discourse on the long-term conditions for economic growth. Hitherto, economic growth has largely been based on using the Earth either as a source of natural resources (like copper and oil) or as a sink for deposits (like CO_2 in the atmosphere). At a global level, this remained a marginal problem as long as growth paths were well within the bounds of what the Earth could accommodate. If that is no longer possible, the potential for economic growth will change (Solow, 1974; Arrow et al., 1995; Simpson et al., 2005; Rockström et al., 2009). Sustainable 'Green growth' has to be based on, and limited to, the energy inflow from the sun. As this is approximately 10,000 times what we presently use (Sandén, 2008), this is—theoretically at least—not a particularly strong limit.

For policy-makers, however, this is a real challenge: the destruction of the black sectors cannot wait. The only solution to achieve net growth is to combine this destruction with rapid creation of a dynamic labour-intensive and green economy based on renewable energy.

References

Abramowitz, M. (1956). 'Resource and Output Trends in the United States since 1870', *American Economic Review, Papers and Proceedings* (May): 5–23; republished in N. Rosenberg (ed.), *The Economics of Technological Change*, Harmondsworth: Penguin, 1971.

Aghion, P., and Howitt, P. (2009). *The Economics of Growth*, Cambridge, Mass.: MIT Press.

Al-Juaied, M., and Whitmore, A. (2009). *Realistic Costs of Carbon Capture*, Discussion Paper 2009–08, Energy Technology Innovation Research Group, Belfer Center for Science and International Affairs. Boston: Harvard Kennedy School, July.

Arrow, K., Bolin, B., and Costanza, R., et al. (1995). 'Economic Growth, Carrying Capacity, and the Environment', *Science*, 268(5210): 520–1.

Azar, C., and Sanden, B. (2011). 'The Elusive Quest for Technology-Neutral Policies', *Environmental Innovation and Societal Transitions*, 1: 135–9.

Berg, M. (1985). *The Age of Manufacures 1700–1820*, Oxford: Basil Blackwell.

Bergek, A., and Jacobsson, S. (2010). 'Are Tradable Green Certificates a Cost-Efficient Policy Driving Technical Change or a Rent Generating Machine? Lessons from Sweden 2003–2008', *Energy Policy*, 38: 1255–71.

BP (2014). *Statistical Review of World Energy* (2014). June <bp.com/statisticalreview>.

Brondel, G. (1976). 'The Sources of Energy', in C. Cipolla, *The Fontana Economic History of Europe*, v, Glasgow: Collins/Fontana Books.

Brynjolfsson, E., and McAfee, A. (2013). *The Second Machine Age*, New York and London: W. W. Norton & Co.

Canadell, J., and Schulze, E.-D. (2014). 'Global Potential of Carbon Biospheric Management for Climate Mitigation', *Nature Communications*, DOI:10.1038/ncomms.

Carbon Tracker (2014). *Carbon Tracker Initiative*, <www.carbontracker.org>.

Castells, M. (2000). *The Rise of the Network Society*, Oxford: Basil Blackwell.

Dahmén, E. (1950). *Svensk Industriell företagarverksamhet*, del 1. Stockholm: Industriens Utredningsinstitut.

Dahmén, E. (1989). 'Development Blocks in Industrial Economics', in B. Carlsson (ed.), *Industrial Dynamics: Technological Organizational and Structural Changes in Industries and Firms*, Boston: Kluwer Academic Publishers, 109–22.

EWEA (2014). 'European Wind Energy Association', <www.ewea.org>.

Erixon, L. (ed.) (2003). *Den svenska modellens ekonomiska politik*, Stockholm: Bokförlaget Atlas.

Esping-Andersen, G. (1990). *The Three Worlds of Welfare Capitalism*, Cambridge: Polity Press.

Eurostat (2014). <http://epp.eurostat.ec.europa.eu/portal/page/portal/statistics/themes>.

Fehske, A., et al. (2011). 'The Global Footprint of Mobile Communications: The Ecological and Economic Perspective', *IEEE Communications Magazine*, Aug.: 55–62.

Finkenrath, M. (2011). *Cost and Performance of Carbon Dioxide Capture from Power Generation*, Paris: OECD/IEA.

Friedman, T. (2005). *The World is Flat: A Brief History of the Globalized World in the 21st Century*, London: Allen Lane/Penguin.

Gardiner, S. (2011). *A Perfect Moral Storm: The Ethical Tragedy of Climate Change*, New York: Oxford University Press.

Garud, R., and Karnoe, P. (eds) (2001). *Path Dependence and Creation*, Mahwah, NJ: Lawrence Erlbaum Associates.

GCCSI (2011). *The Costs of CCS and Other Low-Carbon Technologies*, Melbourne: Global CCS Institute.

Giddens, A. (2009). *The Politics of Climate Change*, London: Polity Press.

Gordon, R. (2012). *Is US Economic Growth Over? Faltering Innovation Confronts the Six Headwinds*, Working Paper, 18315, Cambridge, Mass.: NBER.

Hansen, J. et al. (2008). 'Target Atmospheric CO2: Where Should Humanity Aim?', *The Open Atmospheric Science Journal*, 2: 217–31.

Hansen, J., Sato, M., and Ruedi, R. (2012). 'Perception of Climate Change', *Proceedings of the National Academy of Science*, PNAS Early Edition, <www.pnas.org/cgi/doi/10.1073/pnas> (1205276109).

Hilson, M. (2008). *The Nordic Model: Scandinavia since 1945*, London: Reaction Books.

Hoel, M. (2012). *Klimatpolitik och ledarskap—vilken roll kan ett litet land spela?'*, Rapport till Expertgruppen för miljöstudier, 2012/3, Stockholm: Regeringskansliet, Finansdepartementet.

ICO2N (2014). 'Capture Technology Supply Curve', <www.ico2n.com>.

IEA (2010). *Power Generation from Coal. Measuring and Reporting Efficiency Performance and CO_2 Emissions*, Paris: OECD/IEA.
IEA (2013a). *Redrawing the Energy-Climate Map*, Paris: OECD/IEA.
IEA (2013b). *World Energy Outlook 2013*, Paris: OECD/IEA.
IEA (2014a). *CO_2 Emissions from Fuel Combustion, Highlights*, Paris: OECD/IEA.
IEA (2014b). *Energy Technology Perspectives 2014*, Paris: IEA.
IEA (2014c). *Key World Energy Statistics 2014*, Paris: OECD/IEA.
IEA (2014d). *World Energy Outlook 2014*, Paris: OECD/IEA.
IPCC (2005). *Carbon Dioxide Capture and Storage*, Cambridge: Cambridge University Press.
IPCC (2013). *Climate Change 2013: The Physical Science Basis*, WG1 contribution to AR5, summary volume <www.climatechange2013.org>.
IPCC (2014a). *Climate Change 2014: Mitigation of Climate Change*, WGIII contribution to AR5, <www.mitigation2014.org>.
IPCC (2014b). *Climate Change 2014: Synthesis Report*, <www.ipcc.ch>.
Kline, S., and Rosenberg, N. (1986). 'An Overview of Innovation', in R. Landau and N. Rosenberg (eds), *The Positive Sum Strategy*, Washington, DC: National Academy Press, 275–306.
Kost, C., Mayer, J., et al. (2013). *Levelized Cost of Electricity Renewable Energy Technologies*, Nov., Freiburg: Fraunhofer Institute for Solar Energy Systems.
Laestadius, S. (2013a). *Klimatet och välfärden—mot en ny svensk modell*, Umeå: Borea bokförlag.
Laestadius, S. (2013b). 'Industrial Transformation and Climate Change: The Role of Low-Tech Industries', in Jörg Abel, Gerd Bender, and Katrin Hahn (eds), *Traditionell Innovative*: Festschrift fur Hartmut Hirsch-Kreinsen zum 65. Geburtstag, Berlin: edition sigma.
Landes, D. (1969). *The Unbound Prometheus*, Cambridge: Cambridge University Press.
Larsson, A., et.al. (2009). *An International Climate Investment Community: Breaking the Deadlock*, Stockholm: Global Utmaning.
Lundberg, E. (1985). 'The Rise and Fall of the Swedish Model', *Journal of Economic Literature*, 23 (Mar.): 1–36.
McKinsey (2008a). *Carbon Capture and Storage: Assessing the Economics*, New York: McKinsey Global Institute.
McKinsey (2008b). *The Case for Investing in Energy Productivity*, New York: McKinsey Global Institute.
Maddison, A. (2005). *Growth and Interaction in the World Economy: The Roots of Modernity*, Washington, DC: AEI Press.
Maddison, A. (2007). *Contours of the World Economy 1–2030 AD: Essays in Macro-Economic History*, Oxford: Oxford University Press.
Malmodin, J., et al. (2010). 'Greenhouse Gas Emissions and Operational Electricity Use in the ICT and Entertainment & Media Sectors', *Journal of Industrial Ecology*, 14(5): 770–90.
Markusen, A. (1996). 'Sticky Places in a Slippery Space: A Typology of Industrial Districts', *Economic Geography*, 72: 293–313.
Marshall, A. (1890/1990). *Principles of Economics*, 8th edn, London: Macmillan.

Mokyr, J. (2009). *The Enlightened Economy*, New Haven: Yale University Press.
MottMacdonald (2012). *Potential Cost Reductions in CCS in the Power Sector*, Discussion Paper, London: Dept of Energy and Climate Change.
Nordhaus, W. (2007). 'A Review of the Stern Review on the Economics of Climate Change', *Journal of Economic Literature*, 45: 686–702.
Nordhaus, W. (2011). *Estimates of the Social Cost of Carbon: Background and Results from the RICE-2011 Model*, Cowles Foundation Discussion Paper, 1826. New Haven: Yale University Press.
Nordhaus, W. (2013). *The Climate Casino: Risk, Uncertainty and Economics for a Warming World*, New Haven and London: Yale University Press.
Norgaard, K. M. (2011). *Living in Denial: Climate Change, Emotions, and Everyday Life*, Cambridge, Mass.: MIT Press.
Novotny, M., and Laestadius, S. (2014). 'Beyond Papermaking: Technology and Market-Shift for Wood-Based Biomass Industries—Management Implications for Large-Scale Industries', *Technology Analysis and Strategic Management*, 26(8): 875–91.
Nye, D. (1998). *Consuming Power: A Social History of American Energies*, Cambridge, Mass.: MIT Press.
OECD (2011). *Towards Green Growth*, Paris: OECD. <www.oecd.org/greengrowth/48224539.pdf>.
Ostrom, E. (1990). *Governing the Commons: The Evolution of Institutions for Collective Action*, New York: Cambridge University Press.
Ostrom, E., et al. (2002). *The Drama of the Commons*, Washington, DC: National Academy Press.
Parry, I., Veung, C., and Heine, D. (2014). *How Much Carbon Pricing is in Countries´ own Interests? The Critical Role of Co-Benefits*, IMF Working Paper, WP/14/174, Washington, DC: IMF.
Piketty, T. (2013). *Capital in the Twenty-First Century*, Cambridge, Mass.: Belknap Press.
Pigou, A. C. (1920). *The Economics of Welfare*, London: Macmillan.
Pindyck, R. (2012). *The Climate Policy Dilemma*, NBER Working Paper, 18205, July. Cambridge, Mass.: National Bureau of Economic Research.
Porter, M., and van der Linde, C. (1995). 'Toward a New Conception of the Environment: Competitiveness Relationship', *Journal of Economic Perspectives*, 9(4): 97–118.
Rockström, J., et al. (2009). 'Planetary Boundaries: Exploring the Safe Operating Space for Humanity', *Ecology and Society*, Sept., <www.ecologyandsociety.org/vol14/iss2/art32>, accessed Dec. 2014.
Running, S. (2012). 'A Measurable Planetary Boundary for the Biosphere', *Science*, 337: 1458–9.
Sandén, B. (2008). 'Solar Solution: The Next Industrial Revolution', *Materials Today*, 11 (2): 22–4.
Simpson, D., Toman, M., and Ayres, R. (eds) (2005). *Scarcity and Growth Revisited: Natural Resources and the Environment in the New Millennium*, Washington, DC: Resources for the Future Press.
Smil, V. (2010). *Energy Transitions: History, Requirements Prospects*, Santa Barbara, Calif.: Praeger.

Smith, W. K., Zhao, M., and Running, S. (2014). 'Global Bioenergy Capacity as Constrained by Observed Biospheric Productivity Rates', *Bioscience*, 62: 911–22.

Solow, R. (1957). 'Technical Change and the Aggregate Production Function', *Review of Economics and Statistics*, Aug.: 312–20.

Solow, R. (1974). 'The Economics of Resources and the Resources of Economics', *American Economic Review*, 64(2): 1–14.

Stern, N. (ed.) (2007). *The Economics of Climate Change*, Cambridge: Cambridge University Press.

Stern, N. (2008). 'Upside of a Downturn', interview in *Financial Times* (with Fiona Harvey), 2 Dec., 'Climate Change Part Three: Business'.

Stiglitz, J. (2014). *Unemployment and Innovation*, NBER Working Paper, 20670, Washington, DC: National Bureau of Economic Research.

Tol, R. (2009). 'The Economic Effects of Climate Change', *Journal of Economic Perspectives*, 23(2): 29–51.

Toulmin, S. (1990). *Cosmopolis: The Hidden Agenda of Modernity*, New York: Free Press.

UNECE/FAO (2009 and 2011). *European Forest Sector Outlook Study*, Geneva: Geneva Timber and Forest Study.

UNEP (2014). *The Emissions Gap Report: A UNEP Synthesis Report*, Nairobi: United Nations Environmental Program.

USGS (2014). US Geological Survey, <www.usgs.gov>.

Urry, J. (2011). *Climate Change and Society*, Cambridge: Polity Press.

Van den Bergh. J. C. J. M., and Botzen, W. J. W. (2014). 'A Lower Bound to the Social Cost of CO_2 Emissions', *Nature Climate Change*, 4: 253–8.

Van der Wee, H. (1987). *Prosperity and Upheaval: The World Economy 1945–80*, Harmondsworth: Penguin, Pelican Books.

Von Tunzelmann, N. (1978). *Steam Power and British Industrialization to 1860*, Oxford: Oxford University Press.

Watson, J. (ed.) (2012). *Carbon Capture and Storage: Realising the Potential*, Oxford: UK Energy Research Centre, Apr.

World Bank (2012). *Turn Down the Heat: Why a 4°C Warmer World Must be Avoided*, Washington, DC: World Bank.

Part III

Policies for a Sustainable Europe

7

Lessons from Germany's *Energiewende*

Volkmar Lauber and Staffan Jacobsson

7.1. Introduction

The focus of this chapter is on the triple challenge in one particular sector, namely the electricity-generating sector. In Europe, an almost carbon-free electricity system needs to be in place no later than 2050 (EC, 2011), implying that new capacity delivering 3,500 TWh may need to be built. This is equal to the *total electricity production* in the EU in 2010 (plus Norway and Switzerland) (Box 7.1). A transformation of this scale necessitates long-term and multi-dimensional policies supporting innovation and the diffusion of a range of technologies, a diffusion which involves nothing less than a Schumpeterian process of 'creative destruction'. As in all such processes, opportunities are generated for creative entrepreneurs and policy-makers to open up new areas of growth in a globalized economy. Yet the 'destruction' element implies that a great deal of resistance is to be expected.

The German vision of an energy transition—*Energiewende*—has a history going back to the 1980s. It was first concerned with the phasing out of nuclear power but soon became preoccupied also with energy efficiency and renewable energy sources. It was a transformative project in the direction of a sustainable economy and was soon propelled by strong popular support, while many in the political elite long remained sceptical—and became so again more recently.

Energiewende in the electricity sector includes the development of new sources of growth in the form of capital goods industries (e.g. wind turbines and equipment for manufacturing solar cells) and associated value chains (Jacobsson and Lauber, 2006; Blankenberg and Dewald, 2013). Hence, *Energiewende* is a way to meet the triple challenge in the field of energy and exemplifies both the conception of this transformation process and the evolving policy tools needed for managing such a radical and long-term enterprise.

Box 7.1. THE SCALE OF THE TRANSFORMATION CHALLENGE IN THE EU AND GERMANY

European Union

In the EU, including Switzerland and Norway, electricity production was 3310 TWh in 2001 and 3560 TWh in 2010 (Eurostat, 2012). With the same growth rate (0.85 per cent) henceforth, power supply would be nearly 5000 TWh by 2050. Assuming an unchanged supply of nuclear power (922 TWh in 2009) and full substitution of fossil fuels (1676 TWh), new 'low-carbon' capacity generating over 3100 TWh will be needed by 2050.

Yet, the scale of the transformation is even larger since European nuclear power plants are ageing and few are being built. In the EC (2011) Energy Roadmap 2050, the average nuclear power supply in five decarbonization scenarios is 527 TWh. This means that new capacity delivering 3500 TWh by 2050 may be required, which is equal to the *whole* supply in 2010 and 9.3 times *total* UK supply in 2009 (IEA, 2012).

Germany

Although Germany is a pioneer in new renewables, the goal of *Energiewende* is ambitious. In 2012, electricity supply was 630 TWh and domestic demand 607 TWh. Of the supply, 361 TWh (57 per cent) were fossil-fuel based, 100 TWh (16 per cent) nuclear, and 144 TWh (23 per cent) renewable (AGEB, 2013). Assuming that the 25 per cent reduction goal is reached, consumption would be 455 TWh, with renewables accounting for 364 TWh by 2050. However, if electricity consumption instead grows at the same pace as between 2000 and 2012 (0.4 per cent p.a.), the supply from renewables would have to grow to 565 TWh.

Hence, in addition to replacement investments, new capacity is needed to supply 220–421 TWh of electricity from renewables. With a 100 per cent supply of renewables in 2050, these figures would grow to 311–562 TWh.

By 2011, *Energiewende* had three specific goals for the electricity sector: (i) to replace all nuclear power plants by 2022; (ii) to reduce consumption by 25 per cent compared to 2008; and (iii) to reach at least 80 per cent renewables by 2050 at the latest (EEG, 2011; IEA, 2013).

Even though Germany is a pioneer in new renewables, the overall goal of *Energiewende* is ambitious (Box 7.1). Its realization depends not only on ensuring the large-scale diffusion of new technologies, with challenges in terms of such issues as technology, skills, funding, and coordination (see Jacobsson and Karltorp, 2013, for the case of offshore wind power) but it also requires a simultaneous reduction in fossil-fuel- and nuclear-based electricity. The purpose of this chapter is to identify lessons from the German experience, including lessons from the reception of Germany's *Energiewende* by the European Commission. Section 7.2 analyses how *Energiewende* was conceived and brought forward in a political struggle, leading to the design of a unique policy, the Renewable Energy Act (Gesetz für den Vorrang Erneuerbarer Energien, in short Erneuerbare Energien Gesetz or EEG) to induce the rapid diffusion of renewables and the development of new industries. Section 7.3

identifies opponents of EEG and analyses their resistance. Against this background of the politics involved, section 7.4 addresses the opponents' discourse and its focus on the 'unaffordably' high cost of EEG and the associated 'burden' on consumers. Section 7.5 concludes with a discussion of the lessons learnt from this specific case for meeting the triple challenge more generally.

7.2. *Energiewende* and EEG

EEG is the prime tool for realizing *Energiewende*, the roots of which are to be found in various social movements which still shape contemporary opinion. At the core of these movements is a struggle over energy pathways (Lovins, 1976), manifesting itself in conflicts over regulation, from the first Feed-in Law of 1990 and EEG 2000 to subsequent revisions of the law. EEG 2000 was unique in that it introduced—as distinct from technology-neutral policies—highly differentiated technology-specific payments, allowing simultaneous learning to take place in many technologies. We show how EEG took on its particular shape, developed an unexpected dynamism, and impacted on diffusion, innovation, and industrial growth.

Origin, Intention, and Ascent of EEG

In the 1970s, electricity generation became the subject of intense political controversy. The government reacted to the oil crises by promoting the expansion of nuclear and coal power, with the help of subsidies eventually amounting to hundreds of billions of euros (Küchler and Meyer, 2012). An anti-nuclear movement arose in the mid-1970s; scepticism towards nuclear power became widespread and came to represent the majority view after Chernobyl (Jahn, 1992). A broader 'alternative energy' movement emerged, coining the term *Energiewende* in the late 1970s. After Chernobyl, the Greens committed themselves to a shutdown of all nuclear power plants and the Social Democrats to their gradual phasing out. In public opinion and parliament, demands increased to support the creation of a market for renewables.

While Conservative and Liberal[1] government leaders had always opposed such initiatives, a group of Conservative MPs submitted a Members' bill for a feed-in tariff. These MPs were, however, persuaded to accept a government research, development, and demonstration programme for wind and solar power instead. Another Members' bill was more successful as electricity incumbents were busy with absorbing East German utilities after German

[1] Liberal in the German context means FDP (the Free Democratic Party).

reunification. The Feed-in Law was adopted unanimously in 1990 and required utilities to connect and purchase renewable power from installations not owned by government or utilities, via feed-in tariffs set as fixed percentages of household consumer tariffs (at least 90 per cent for wind and solar power; 80 per cent for biomass power; lower rates for the rest). When the incumbents realized the law's impact, they tried unsuccessfully to have it overturned by German courts and EU authorities or to have it replaced by more 'harmless' legislation (Jacobsson and Lauber, 2006; Laird and Stefes, 2009).

In 2000, the new Social Democratic-Green coalition (1998–2005) introduced a more radical law: the Renewable Energy Act (EEG) of 2000. The Explanatory Memorandum attached to it (FME, 2000) demonstrated an awareness of two major issues. First, it emphasized that the relatively high additional costs of renewables were overstated due to subsidies to coal and nuclear generation, and lack of internalization of their external costs. Second, it articulated the belief—then widespread in the renewable energy community—that it was necessary to break the vicious circle of low demand and high prices for capital goods (turbines, solar cells etc.). This could be achieved by producing equipment on an industrial scale, inducing learning, and bringing down costs. The associated capital goods industries would then be strengthened, leading to additional benefits in the form of employment and exports.

For this purpose, the Act created considerable investment security via the obligation on utilities to connect and purchase all renewable electricity generation offered to them and to pay tariffs set by law for twenty years. Tariffs for new installations in subsequent years were to be reduced at a set annual rate (with the cuts reflecting expected learning) which could be adjusted by later amendments. All relevant technologies were supported simultaneously, with tariffs highly differentiated by technology, size, and so on to minimize windfall profits.[2] The cost of the tariffs was to be passed on to all electricity consumers, but the first exceptions for energy-intensive industries facing international competition were made in 2003.

While the 1990 Feed-in Law increased electricity generation from renewables from 3.5 per cent (20 TWh) to about 5 per cent (29 TWh), the authors of EEG 2000 had greater ambitions. Within two decades, they intended to replace nuclear power, which then provided about 30 per cent of supply (170 TWh). Fossil fuels would be replaced later (346 TWh), although the timing for this was less clear. These ambitions, and the tools for reaching

[2] Despite repeated European Commission attempts to eliminate feed-in tariffs, a report by Commission services (European Commission, 2008) found feed-in tariffs to be the most effective and cost-efficient instrument for renewables support. This was echoed by the IEA (IEA, 2008; Lauber, 2011).

them, were controversial. The Conservatives criticized EEG's excessive interference with the market while the FDP (Liberals) rejected the instrument altogether. Both rejected a phasing out of nuclear energy.

The Impact of EEG

The impact of EEG has been significant in terms of deployment, growth of a capital goods industry, and employment. First, it had an unexpectedly large impact on diffusion. Amendments to EEG (in 2004 and 2008) expanded support for solar, biomass, and offshore wind power and set more ambitious targets. The original target for 2010 (12.5 per cent) had already been reached in 2007. When a Conservative-Social Democratic coalition (2005–9) overhauled the Act in 2008, it raised the aim to 'at least 30%' by 2020—the first time that the Conservatives had voted in favour of EEG. Continuing this trend, Germany's National Renewable Energy Action Plan to the European Commission (Bundesregierung, 2010) increased the target to 38.6 per cent by 2020.

By 2013, the total supply of renewable electricity had grown to 152 TWh, out of which the three largest sources were wind (53 TWh), biomass (43 TWh), and solar cells (30 TWh), while the supply of nuclear power had been reduced from 170 to 97 TWh and that of coal from 291 to 286 TWh (see Figure 7.1). The electricity mix was, thus, substantially changed, with renewables' share growing from 6.6 to 23.9 per cent, nuclear being nearly halved to 15 per cent and coal reduced from 50 to 45 per cent. Continuing with average installation

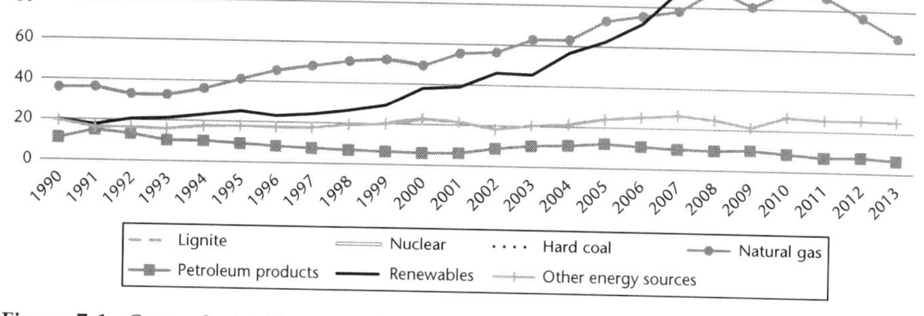

Figure 7.1. Gross electricity generation in Germany by sources, in TWh

levels between 2009 and 2013 for renewables, a two-thirds share by 2030 seemed within reach.[3]

Second, EEG provided early local markets which, together with supplementary policies,[4] opened up growth opportunities for capital goods suppliers. In the value chain for wind turbines (both onshore and offshore), German suppliers abound and include Enercon, Siemens, Nordex, and Repower/Senvion. In solar cells, while German suppliers of cells and modules have recently met with fierce competition from China and declining sales if not insolvency (Blankenberg and Dewald, 2013), VDMA (2013) reports that German companies have about 50 per cent of the world market for solar cell manufacturing equipment (Wessendorf, 2013).[5] Hoppmann et al. (2013) convincingly argue that EEG was conducive to the growth of a specialized equipment industry, and that the availability of embodied technology facilitated fast and easy scaling-up of production in Germany and elsewhere (see also the discussion in Chapter 5 by Schmitz and Lema in this volume).[6]

The development of a German capital goods industry is reflected in a rapid growth in patent applications. In the category of 'renewable energy generation', these grew from 101 in 1999 to 536 in 2009.[7] Germany's share of such applications in EU was 52 per cent in 1999 and it remained high in 2009 (40 per cent). Germany's share of world patent applications in this field was as high as 26 per cent in 1999 and 19 per cent in 2009.[8] Interestingly, the German share remained broadly constant in solar cells—50 per cent of EU and 20 per cent of world applications—but dropped in the case of wind, although it remained at a respectable level—34 per cent of EU applications and 20 per cent of world applications.

Third, employment in the renewable energy sector grew from 100,000 in 2000 to over 350,000 in 2012. Wind, solar, and biomass dominated, with more than 100,000 each (Renewable Energies Agency, 2013).[9]

[3] Based on extrapolation from AGEB (2014) figures and assuming no growth of consumption.

[4] See e.g. Bergek and Jacobsson (2003) and Jacobsson and Bergek (2004).

[5] Turnover in 2011 was about €2.8 billion and the industry had 12,000 employees in 2010. The export ratio was 90% (Wessendorf, 2013). For the first half of 2014, a world market share of over 50% was claimed (VDMA, 2014).

[6] See also Blankenberg and Dewald (2013).

[7] These are the figures for applications to the European Patent Office by inventor's country of residence and by priority date in the category 'Renewable energy generation' (OECD, StatExtracts, patents in environment-related technologies, accessed Dec. 2013).

[8] The level for 2009 is nearly exactly that for all patent classes, which means that Germany has achieved the same strength in renewable energy generation as in industry at large.

[9] However, employment in the German PV sector fell sharply in 2013, in line with the fall in new installations (Solarthemen, 2014).

7.3. Opponents to EEG and *Energiewende*

In this section, we identify opponents of EEG and detail how they contested it with increasing intensity, aiming to slow down or even cap the deployment of renewables. Much of the discourse centres on the alleged 'unaffordably' high cost of renewables, a notion that is addressed in section 7.4.

The Opponents

There are two main opponents. First, large utilities are threatened by the encroachment of renewables in which they have invested only marginally so far (see Box 7.2 for some features of the German electricity market). This threat is magnified since the diffusion of solar and wind power reduces prices and volumes of conventional peak generation from which incumbents used to draw most of their profits; now peak hours for them mean lower sales at lower prices (*The Economist*, 2013). Large utilities are, indeed, reportedly facing an 'existential threat' from renewables, at the same time that they are building new coal-based power plants (IEA, 2013; Rogol, 2011; Becker, 2011) and phasing out redundant plants.

Second, MacKerron (2011) explains that there is a gulf between those inspired by neoclassical economics to favour 'non-discriminatory' policies and those arguing the need for technology-specific policies and, therefore, an active and stronger state. In the German context, the former suggest that the solution lies not in EEG but in 'market-based' quota systems for electricity generation (such as the Swedish-Norwegian tradable green certificate system)

Box 7.2. FEATURES OF THE GERMAN ELECTRICITY MARKET

The German wholesale electricity market is an 'electricity only' market, i.e. the merit order at the exchange is based on marginal costs only, ignoring capacity costs. Thus, solar and wind power (with marginal costs close to zero) rank first; then nuclear, followed by lignite and hard coal plants. As higher cost power plants (oil, followed by gas) are squeezed out and noontime price peaks are levelled due to the near-zero marginal cost of solar generation, wholesale prices of electricity decrease. The 'Big Four' incumbents (RWE, E.ON, EnBW, and Vattenfall), and utilities more generally, own very little renewable generation capacity, but have strongly increased coal capacity recently. Together with growing renewable supplies and declining demand, this has reduced spot market prices by more than half since 2008. This, in turn, benefits big industry which buys its supply directly on the exchange.

EEG grants priority grid access to renewable power. Conventional sources provide 'residual load' which has been in steady decline since 2008. Since 2010 EEG renewable power has been sold on the spot market, while a surcharge covers the difference between spot market price and payments to renewable generators under the feed-in tariffs. This surcharge is levied mostly on small consumers (households and SMEs).

or bidding systems with quotas. Both policy instruments favour the selection of the *currently* most cost-efficient solutions. The latter group of analysts and policy-makers argue that current cost-efficiency constitutes an inadequate selection criterion and emphasize the need for providing learning spaces over long time periods in order to foster new capital goods industries with the ability to deploy new technologies *in time* to meet climate change goals, and at a reasonable cost. We will first analyse the coal-based incumbents, including policy towards them, and then attempts by the Conservative-Liberal (2009–13) and the Conservative-Social Democratic (from 2013 onwards) governments to slow down the diffusion of renewables.

The Coal-Based Incumbents

Fossil fuels have long been subsidized in Germany. With the oil crises of the 1970s, the government required utilities to burn a quota of German-mined hard coal and subsidized them via a surcharge on electricity consumers. This method of financing was held to be unconstitutional in 1995, so funding was transferred to public budgets (of the federal government and governments of coal-mining states) and gradually phased out (supposedly by 2018). Total subsidies were large: for hard coal (and for all programmes from 1970 to 2012) the total came to about €187 billion, and for lignite €65 billion (Storchmann, 2005; Küchler and Meyer, 2012).

Unlike renewable and nuclear policies, coal policy was not so controversial in domestic German politics, despite its high cost.[10] However, the European Commission insisted on reducing such state aid, and electricity liberalization in the late 1990s led to the stagnation of fossil-fuel power plant investment due to surplus capacity and new market risks for an industry little used to competition. From 2001 to 2008, natural gas represented about 70 per cent of new fossil-fuel capacity installed, with lignite over 20 per cent, and hard coal negligible. But in 2008, fossil-fuel plants under construction—some 35 GW—were about 60 per cent hard coal, 20 per cent lignite, and only 20 per cent natural gas. Nearly all coal plants were to be built by the four big incumbents in one of the biggest waves of investment in coal capacity since post-war reconstruction (IEA, 2013).

Why was there this 'dash for coal' in parallel with increasing government ambitions for renewables? Presumably, the incumbents were surprised by the rapid growth of renewable power, their own failure at having EEG overturned, and the stagnation of demand since 2008. Pahle (2010) and Pahle et al. (2011) also point out that the rise in natural gas prices during the preceding years had

[10] Criticizing hard coal subsidies became taboo for politicians in North Rhine-Westphalia, the main coal-mining region and Germany's most populous state.

made coal-fired plants more lucrative than plants fired by gas when environmental effects are not paid for. They also emphasize that this cost-advantage was strengthened by the way the EU Emissions Trading System (ETS)[11] was implemented by Germany. The first German National Allocation Plan of 2005 granted free carbon allowances to existing fossil-fuel power plants according to previous emissions (a process known as 'grandfathering') and to successor plants according to projected future (not actual) emissions. This system favoured coal over gas, since more allowances were granted for coal than for gas plants. In 2005 alone, charging free allowances to customers created estimated windfall profits of €5 billion just for the two biggest incumbents, RWE and E.ON (Pahle 2010), who used the proceeds to finance investments in further coal plants.[12] Such windfalls, of course, fall outside the usual definition of a subsidy. Finally, coal enjoyed domestic government support from unexpected quarters. In 2009, the Environment Ministry, normally the chief institutional advocate of renewables and then led by Social Democrat Sigmar Gabriel, published an energy roadmap targeting 40 per cent of supply from highly efficient—low emission but non-CCS (Box 7.3)—coal power plants by 2020 and assumed a 50 per cent CCS retrofit for all coal plants by 2030 (BMU, 2009). Given their lifespan of half a century, this implied a future for coal plants well beyond 2050.[13]

Hence, until 2008, German coal generators did well but from then on the situation changed. First, declining wholesale prices drove down profits, largely due to the unexpectedly rapid surge of renewables and the 'merit-order effect', which means that the least cost-effective plants (in terms of marginal costs)

Box 7.3. CARBON CAPTURE AND STORAGE (CCS) IN GERMANY

Coal power still accounts for a large share of electricity supply and Germany has vast reserves of lignite, which make carbon capture and storage technologies attractive. The Swedish utility Vattenfall began to operate a 30 MW demonstration plant in Schwarze Pumpe in 2008 and the Energy Concept of 2010 recognized the potential role of CCS in the power and energy-intensive industrial sectors. Public resistance constitutes a major barrier, however, and CCS has met with great reservations in those areas deemed suitable for CO_2 storage. The CCS Act of 2012 was highly controversial and only allows for CCS on a test basis, restricts the amount of CO_2 to be stored to 1.3 Mt/year per storage site and allows individual *Länder* to exclude parts of the state territory from CO_2 storage (IEA, 2013).

[11] For a description of ETS, see Box 8.4 in this volume.

[12] Large coal plants cost about €1 billion (Pahle, 2010).

[13] In addition, high credit ratings and substantial cash flows of the four big incumbents meant low-cost financing; as did a strong increase in wholesale electricity prices—by about 80% between 2000 and 2007.

are no longer dispatched (Box 7.1) but replaced by renewables. Since renewables and nuclear energy have very low marginal costs, an increase in the supply of renewables will *ceteris paribus* lead to reduced sales for plants driven by fossil fuels. Gas plants with their higher operating costs were hit first; hard coal plants are also claimed to run fewer full load hours, and peak supply—a lucrative market previously—needs much less fossil power with the widespread diffusion of wind turbines and solar cells (as noted, it was during daily peak hours that hard coal generators made most of their profits). Second, while all demand contracted with the economic crisis, demand for conventional power declined additionally with the rise of renewables.[14] All this led to surplus capacity, intensified competition, and more profit falls. Third, windfall profits from ETS for new plants disappeared in 2013, since plants from then on had to buy their certificates in auctions. Fourth, coal subsidies directed specifically at generators are scheduled to expire in 2018 (IEA, 2013). Fifth, in addition to problems with incumbents' coal plants, upcoming nuclear decommissioning is hurting them to the point that three of them offered to transfer their nuclear plants to a government foundation (which would run them until 2022), adding some €35 billion set aside for decommissioning plants and storing nuclear waste, and to drop their lawsuit for damages resulting from the nuclear phase-out decision of 2011. In exchange, they wanted to be relieved of their current liability for decommissioning and storage (Spiegel, 2014).

All this means declining profits (or even losses) for fossil-fuel power plants and falling stock values for incumbents, which translate into poor credit ratings and higher costs of capital (*The Economist*, 2013). Slowing down *Energiewende* would ease their pain. Making renewables less attractive by reducing feed-in tariffs, capping deployment, or weakening priority grid access for renewable generation would serve this purpose. It is, therefore, not surprising that incumbents should be calling for help. As the situation of coal generators deteriorated, the appeals became more dramatic. The deployment of renewables under EEG is, therefore, presented as not only risking blackouts[15] but also as being 'unaffordable', creating hardship through energy poverty[16] and endangering German industry's competitiveness, exports, employment, and prosperity, in striking contrast with current facts.[17] Only high-level politics

[14] Demand peaked at 622 TWh in 2007 but then declined to 600 TWh in 2013; a strong decline was expected for 2014.

[15] This has been a recurrent theme of the incumbents since 2004, ignoring the fact that Germany has the best record of grid stability in Europe, something that has remained unchanged by renewables' growth.

[16] This criticism mainly comes from organizations not previously known for their concern for the poor, while the usual speakers for the poor tend to see things differently (Morris, 2013).

[17] About half of German industry was exempted from the EEG surcharge (until Aug. 2014), the electricity tax and grid fees, and actually benefits from the low wholesale prices of electricity resulting from merit-order effects.

could bring about a change that would slow down *Energiewende*, which takes us to the second set of challengers.

Policy of the Conservative-Liberal Government 2009–13

The first policy of the new government was to delay the nuclear phase-out to allow renewables to become more 'affordable' in the meantime.[18] Anti-nuclear feeling after the Fukushima catastrophe led, however, to a reinstatement of the phase-out and to declarations that *Energiewende*—which now entered Conservative vocabulary—would be pursued 'even more vigorously'. But the schedule for the deployment of renewables first set out in 2010 was only marginally accelerated in EEG 2011.[19]

Attention was instead focused on solar cells, which then received very high feed-in tariffs. The Conservatives' business wing and the Liberals (FDP) opposed open-ended market creation for solar cells (i.e. without a stringent cap). Even in the ecological wing of the Conservatives, the view prevailed that renewables needed more exposure to competition to reduce prices faster. In early 2010, when new solar cell installations were forecast to more than double (after doubling in 2008 and 2009), Environment Minister Röttgen responded by submitting an amendment to EEG containing a tariff reduction plus a 'flexible cap' for solar cells, automatically reducing tariffs beyond the scheduled decreases if installations exceeded 3.5 GW per year. Such an instrument had been proposed in 2008 by both Pfeiffer (from the Conservative business wing) and Fell (Greens), but was then rejected by the coalition government dominated by the Conservatives' ecological wing and the Social Democrats. Röttgen's reform brought down tariffs but not installations (they reached about 7.5 GW in 2010, 2011, and 2012), the cost of which declined even faster than the tariffs. The Conservatives' business wing and the Liberals were dissatisfied and argued that 'overpriced' tariffs had to come down much more, demanding a steep one-time tariff cut plus a faster decline schedule.[20]

When it became clear in early 2012 that Röttgen's amendment had failed to contain solar cell diffusion (7.5 GW in 2011), he was told to come to terms with Liberal Economic Affairs Minister Rösler, his fiercest opponent and EEG critic. Together, they prepared a bill with tighter, steeply declining flexible caps for solar cells, designed to bring down installations to 0.9–1.9 GW/a by 2017.[21] The opposition parties accused the government of abandoning

[18] See also Lauber and Buschmann (2013).

[19] The targets were 35% by 2020, 50% by 2030, 65% by 2040, and 80% by 2050. Adding 'at least' and 'at the latest' to the target formulation made clear that these targets were not intended as caps.

[20] The fact that most solar cells were imported from China weakened the case of EEG supporters.

[21] Very low (flexible) caps are bound to be exceeded, and in that case flexibility operates to further accelerate tariff reductions.

Energiewende and intentionally destroying the German solar cell industry, and a two-thirds majority of the upper house of parliament, in which the *Länder* governments are represented, refused to accept the bill.[22] A conciliation committee produced a milder version of the bill which set a total cap at 52 GW for EEG support of PV (about twice the capacity installed at the time) and reinstated the flexible target of 3.5 GW annually.

The Liberal leadership called the outcome of the conciliation committee a betrayal of the market economy and attacked EEG principles (see section 7.2) head-on: its obligation on utilities/grid operators to purchase all renewable electricity; the fixed twenty-year tariffs giving security to investors; tariff differentiation allowing the simultaneous deployment of technologies with different cost levels; and finally, priority access to the grid. It also demanded a moratorium on new installations until EEG was reformed.

Finally, it tried to force the hand of the Conservatives by contriving the 'EEG surcharge crisis' of 2012–13, dramatizing the increase of the surcharge to be paid by consumers for 2013 (by almost 50 per cent to about 5.2 Ct/kWh) and linking it to increased feed-in tariff payments. This surcharge supposedly reflects the extra cost of feed-in tariffs by paying grid operators for the difference between the tariffs they pay to EEG generators and the spot market price of the electricity they obtain when they sell it on the exchange.[23]

However, as already pointed out, the rising share of renewables had eliminated the more expensive fossil generators (the 'merit-order effect'), bringing down wholesale prices. These low prices benefit big industrial clients (who buy their electricity on the exchange and are exempted from the surcharge)[24] but not households and small industry, unless the suppliers pass reduced exchange prices on to them, something they have failed to do so far (BMU, 2013). This means that households and other small consumers, such as small- and medium-sized businesses, pay for the exemption of the big industrial consumers while not benefiting from the merit-order effect. Indeed, about half of the surcharge increase for small consumers since 2000 does not reflect additional costs for new renewables installations, but rather growing industry exemptions and falling wholesale prices on the spot market (Mayer and Burger, 2014).

[22] Some Conservative-ruled *Länder* joined in the rejection.

[23] In 2009, the Conservative-Social Democratic government changed the marketing of EEG electricity. Up until 2010, the electricity was physically passed on to distributors and received prices usual at that level. Since 2010 it had to be sold on the spot market, which led to a marked fall in its price (by over a third) and a consequent addition of €1.8 billion to the EEG surcharge in just one year.

[24] Industry exemptions were introduced early but Liberal Economic Affairs Minister Rösler increased these greatly, multiplying their annual volume from one to about four billion in the space of just three years (Mayer and Burger, 2014, fig. 12; IWR, 2014).

However, the political debate initiated by Rösler—and taken up by new Environment Minister Altmaier—ignored all this and focused instead on the 'affordability' of renewables in terms of energy poverty and the supposed dangers to German industry (half of which actually benefits from the combination of surcharge exemptions and record-low wholesale prices—see Küchler and Wronski, 2014). Altmaier proposed capping the surcharge for two years, lowering feed-in tariffs (for all technologies), and reducing industry exemptions. He also proposed postponing feed-in payments to newly connected generators for several months, temporarily limiting their revenues to sales on the spot market. Once more, the initiative failed due to *Länder* opposition but did have the effect of turning off investors, for example, for offshore wind, as will be discussed in section 7.4.

Policy of the Conservative-Social Democratic Government (2013 onwards)

The coalition agreement signed after the 2013 elections (CDU, CSU, and SPD, 2013) showed that the Social Democrat leadership, in a weak position after the election,[25] had changed their stance on renewable energy beyond recognition. The 'coal faction' had gained the upper hand, reflected in the choice of Hannelore Kraft as chief SPD negotiator for energy issues; she was governor of the coal state, North Rhine Westphalia, and an outspoken supporter of new coal power plants. The policy advocated with little success by Altmaier was now mirrored in the coalition agreement, which included the following items: flexible caps and 'corridors' for all renewable power (i.e. trajectories for deployment with minimum and maximum target values); reduced compensation for onshore wind; slower deployment for offshore wind (the 2030 target was reduced from 25 to 15 GW); drastic reductions for new biomass capacity; replacement of feed-in tariffs by market premiums and, in 2018, by a tender system; and an obligation for all consumers of self-generated electricity to contribute to the EEG surcharge.

This policy clearly favours incumbents (with a slower phase-out for coal plants, and bidding systems which favour incumbents for new renewables deployment) while making renewables less attractive. It signals that the advocates of slowing renewables and 'saving' coal (mostly lignite) and coal jobs,[26] allied with big industrial consumers arguing that the high cost of renewable power endangered the competitiveness of German industry as the basis of wealth and employment, could now count on the support of the Social Democratic leadership. In fact, in subsequent months Sigmar Gabriel, Social Democratic Minister of Economic Affairs and Energy, went further in the

[25] CDU-CSU received 41.5% of the vote and SPD 25.7%. In 2005–9 they were almost even.
[26] Employment in coal mining declined from 750,000 in the 1950s to some 30,000 in 2013.

favourable treatment of incumbents and in rejecting EEG philosophy. He brought forward the introduction of tender systems to 2017, tightened caps on solar-PV, and announced the full-scale replacement of EEG by 2016. By contrast, an overwhelming share of the German population—including most Conservative and Social Democratic voters (EMNID, 2013; AEE, 2014)—remained committed to EEG-style *Energiewende* by 'citizen investments' relying more on decentralized generation than on utility-scale plants.

7.4. A Critical Analysis of the Opponents' Cost Discourse

The opponents of EEG, and by implication of *Energiewende*, have always been present but the efforts to delegitimize EEG gained considerable momentum with the declining fortunes of coal and a simultaneous surge of 'more expensive' renewables. We will address the cost discourse in which the two sets of opponents joined forces, and highlight the confusion between different cost concepts ('private', 'consumer', and 'social' costs) and the associated problems of the terms 'cost-efficiency', 'subsidies', and 'affordability'. We will also explain the importance of avoiding short-term consumer or private costs as selection criteria when the objective is to stimulate the formation of innovative capital goods industries with the capacity to transform the electricity system in a few decades. This involves engagement with two themes that are persistent in the German debate: (i) the timescale required for a transformation, and (ii) the role of market formation in the innovation process.

Confusion of Cost Concepts and What Cost-Efficiency Entails

There is a great deal of confusion over what the term 'cost' entails and, therefore, the meaning of cost-efficiency and, consequently, subsidies. A first concept is that of *private*, i.e. the firm's, generation costs. A second is *consumer* costs, which is the cost concept currently focused on by EEG critics (e.g. Frondel et al., 2010). There is often a confusion of these two concepts in that they are believed to be synonymous, but as the former neglects rents, which are included in consumer costs, they are quite distinct (Bergek and Jacobsson, 2010; del Rio and Cerdá, 2014).

Third, a focus on private and consumer costs means that significant cost items are left out of the analysis. The first are external costs,[27] which are those that electricity suppliers and users impose on others without paying for the

[27] Although this is a well-established concept in economic theory, some economists, like Frondel et al. (2010), neglect the role of externalities in this cost discussion, as do many other contributors to the debate.

consequences. These are real in the sense that they are paid by those who suffer, for example, from respiratory diseases and various effects of climate change, such as the flooding of coastal cities (or the cost of avoiding it), droughts, and extreme weather events. A second item is subsidies from (federal and state) government. Externalities, subsidies, and private costs are added together to arrive at the *social cost*. Cost-efficiency then means that social costs are minimized, and in the context of energy policy, it is these costs that are most relevant. In what follows, we will attempt to compare the social cost of coal-generated electricity to that of renewable power.

The German Federal Environment Agency (UBA, 2012) estimates *external costs* to be 10.75 cents/kWh for lignite and 8.94 cents/kWh for hard coal. These figures are based on 'best-practice' cost estimates of climate change effects and other damages from emissions[28] using, for example, a cost estimate of $€80_{2010}$/t CO_2. UBA (2008) recognizes the methodological difficulties in estimating these costs as well as the ethical dimension involved; for instance, what is the value of a life lost and how much consideration should be given to future generations (via the choice of discount rate)? In making these estimates, the methodological problems were tackled and the value judgements addressed in a dialogue with other, largely German, organizations.[29] Although estimates of external costs may be controversial, it is appropriate to use those carried out by a German Government body in a study of Germany, in particular when the estimates do not differ much from the ones in a recent major study commissioned by the European Commission (Alberici et al., 2014, table A3-8).[30]

Küchler and Meyer (2012) provide estimates of the *financial subsidies* channelled through government budgets.[31] *Private costs* for coal are taken from Fraunhofer (2013).[32] The cost of renewables is estimated by adding the feed-in remuneration paid in 2012 per technology—as an approximation of *private*

[28] Climate change costs dominate. For lignite, these are estimated to be 8.68 eurocents/kWh and the costs of other emissions come to 2.07 eurocents/kWh (UBA, 2012).

[29] In 2007, Umweltbundesamt (UBA) prepared a convention on the methodology for estimating external costs after consulting a large variety of government agencies and experts. This was updated in 2012 after additional consultations with INFRAS (Zürich), Fraunhofer ISI, Energy Research Institute IER (Stuttgart), and others (see UBA, 2012).

[30] The estimates are 9.5 eurocents/kWh for hard coal, 8.2 for lignite, 1.4 for solar, 0.4 for onshore wind, and 0.1 for hydro.

[31] The figures for subsidies are taken from appendix tables 3, 4, and 6 in Küchler and Meyer (2012). We include government subsidies (Staatliche Förderungen) in the form of (i) financial aid for all and (ii) exemptions from mining royalties and water levies for coal. The source does not provide figures for each type of renewables, only aggregate figures, which is why they all are subsidized by 0.4 eurocents/kWh in Table 7.1. The figures for subsidies are then divided by the supply of electricity in 2012 (AGEB, 2013). There may be other subsidies which are not captured by these data. We are grateful to Swantje Küchler for help with these estimates.

[32] We use the average of the ranges given in Fraunhofer (2013, 17) but deduct the estimated cost for ETS. This cost was calculated using CO_2 emissions (in kg/kWh) as given in Fraunhofer (2013, table 2) and the cost of CO_2 certificates. For the latter, we use a figure of €5.3/t CO_2.

costs—to estimates of external costs and financial subsidies through the state budget, as provided in UBA (2012) and Küchler and Meyer (2012) respectively. It is important to appreciate that, for renewables, the appropriate term is not current costs but *legacy* costs, since the feed-in cost in 2012 constitutes an average of payments to earlier installations with higher tariffs/kWh, and payments to new installations with lower tariffs/kWh.

Table 7.1 contains the estimates of private costs, external costs, subsidies, and social costs for onshore wind, hydro, PV, and coal. The *private* cost for coal is clearly much below the *private* cost of renewables. This is in line with the message from other analyses that fail to include external costs and subsidies. For instance, Frondel et al. (2010, 4049) argue that:

> utilities are obliged to accept the delivery of power...into their own grid...paying...feed–in tariffs far above their own production costs...[E]ven on-shore wind...requires feed-in tariffs that exceed the per kWh cost of conventional electricity by up to 300% to remain competitive.

However, it is plain that the *social* cost of onshore wind and hydro is much below that of coal. Indeed, in spite of the high legacy cost for PV, the *weighted average social* cost/kWh for the present stock of onshore wind, hydro, and solar cells is the same as for coal-generation facilities—about 17 eurocents/kWh.[33] Thus, the 'burden' of renewables disappears if social costs are used instead of private costs.

As importantly, and as can be gleaned from Table 7.1, the historically high feed-in rates for solar cells have a large impact on the weighted average social cost for the three renewables but *current* feed-in rates (9.5 to 13.7 cents in January 2014) are much lower than in the past (Solarförderverein, 2013). As external costs and subsidies are low (Küchler and Meyer, 2012, appendix table 8), even solar cells are now competitive with coal, which implies that, henceforth, it is *coal power which is the inefficient option in terms of social cost.*

Table 7.1. An estimate of the private and social cost of electricity supply from renewables and coal in Germany in 2012, in eurocents/kWh

Technology	Private costs	External costs	Subsidies through state budget	Social costs
Onshore wind	8.8	0.26	0.4	9.46
Hydro	8.5	0.18	0.4	9.08
PV	36.5	1.18	0.4	38.08
Hard coal	7.0	8.94	1.8	17.74
Lignite	4.4	10.75	1.2	16.35

Sources: UBA (2012); Küchler and Meyer (2012, appendix table 8); AGEB (2013), and Fraunhofer (2013).

[33] The supply of electricity from the respective technologies used in this calculation was 51.0 TWh for wind, 21.8 for hydro, 26.4 for PV, 116.4 for hard coal, and 160.7 for lignite.

In order to capture how the recent dramatic reduction in the cost of solar cells impacts on the *weighted average social cost* of the three renewables, we have replaced the private legacy cost for just solar cells in Table 7.1 with the cost in the middle of the range (11.6) of feed-in tariffs in January 2014. To this estimate of private cost, we have added 1.2 cents in external costs and 0.4 cents in subsidies for solar cells. This generates a weighted average social cost of 10.4 eurocents for the three renewables, which is substantially below that of coal power (17.0).[34]

In summary, the cost discourse gives a misleading image of the supposedly heavy 'burden' of renewables—and of *Energiewende*—and its neglect of important cost items raises strong doubts about arguments referring to 'cost-efficiency' and 'subsidies'. Moreover, it obscures the social cost efficiency of renewables and, therefore, offers poor guidance to future policies.

Innovation and Development of a Capital Goods Industry

An innovative capital goods industry is essential for providing a rapidly rising volume of 'low-carbon' electricity, at reasonable costs, as other energy sources are phased out in the second quarter of this century. The shift in focus from long-term benefits, primarily in the form of avoiding climate change, to short-term consumer costs means that the discourse has come to play down the role of the capital goods industry which, as MacKerron (2011) argues, is the bridge between policy, market formation, and technical change.

The inherently long time-scale for the development and diffusion of new technologies has been well documented and spans several decades (e.g. Grübler, 1996; Wilson, 2012). In the innovation system literature, efforts have been put into assessing the length and character of the 'formative phase' in which the technology is 'put on the shelf', in other words, a rudimentary capital goods industry is developed that provides a technology with a reasonable price/performance (e.g. Jacobsson and Bergek, 2004; Suurs, 2009; Hellsmark, 2010). This phase often takes a couple of decades and an additional two to three decades may be required to increase the capacity of the capital goods sector and deploy the technology (in improved forms) until the market is saturated.

Onshore wind and solar cells have gone through the formative phase and can be deployed on a large scale with social costs lower than for coal, backed by an innovative capital goods sector. The 1991 feed-in law and EEG greatly contributed to the maturing of these two technologies. Offshore wind power is currently in its formative phase (Jacobsson and Karltorp, 2013), but is vital

[34] When greenhouse gases accumulate, the climate cost of fossil-fuel use will rise. UBA (2012) estimates that by 2030 this will increase from the current figure of €80/t to €145/t.

Box 7.4. THE NEED FOR UP-SCALING OF THE CAPITAL GOODS INDUSTRY FOR OFFSHORE WIND POWER

A simple scenario may illustrate the magnitude and timescale involved in strengthening the capital goods industry and diffusing offshore wind power on a large scale in Europe. The annual installed new capacity of offshore wind turbines was 1.2 GW in 2012 and is estimated to increase to 1.9 GW in 2014 (EWEA, 2013). If we are to reach the EU targets of a capacity of 44 GW in 2020 and about 230 GW by 2050, the supply capacity needs to grow by 22 per cent per year until 2022 when it would reach 9.3 GW p.a. Maintaining that supply until 2050 would lead to a cumulative capacity of 301 GW, from which we need to draw replacement investments: assuming twenty-five years' lifetime, the stock would be 223 GW by 2050. Hence, a very significant up-scaling is required in order to even come close to the EU targets.

for the replacement of fossil fuels and needs to shift into a growth phase and, ultimately, mass deployment. In Germany, fossil fuel and nuclear energy generated 457 TWh in 2013; offshore wind power is, arguably, a key technology for their replacement. Until 2013, the German goal was to build a capacity of 25 GW by 2030 (E.ON, 2011), delivering perhaps 87 TWh.[35] In the longer run, much higher figures are expected, both in Germany (German Advisory Council on the Environment, 2011) and the EU, which requires a significant scaling-up of the capital goods industry (Box 7.4).

A Northern European supply chain is, indeed, being developed and Germany is integral to this effort, both as a market and a supplier of capital goods and services. Danish and German firms dominate the capital goods industry and large investments are made in the value chain, including harbours, to develop a complete supply chain. As mentioned in section 7.3, the proposal by Environment Minister Altmaier in early 2013 to cap EEG payments led, however, to major political uncertainty, making investors hesitant. As the German Association of Energy and Water Industries (BDEW, 2013, 21) argued:

> the discussion surrounding the brake on electricity prices . . . has led to further uncertainty amongst investors. As a result, the originally expected expansion of offshore wind farms . . . will be delayed—as will, therefore, the associated cost reduction effects.

The subsequent reduction of the 2030 target from 25 to 15 GW in the 2013 coalition agreement (see section 7.3) delays the expansion further, endangering the formation of a supply chain large enough to develop and diffuse offshore wind turbines on the required scale, in time, and at a reasonable cost.

[35] A 40% capacity factor is assumed.

With the long time-scales involved (and the associated learning processes), current investments should not be judged solely by their present costs but also by their ability to contribute to avoiding future costs through enabling the development of an innovative supply chain. An appropriate cost concept should, therefore, include long-term benefits from learning, further strengthening the economic case of renewables, as was the philosophy behind EEG 2000 (FME, 2000).

FDP, Frondel et al. (2010), the German Council of Economic Experts (Sachverständigenrat, 2012), and the Monopoly Commission (2013) all insisted, however, on focusing on short-term costs and argued that the best solution to reduce the alleged 'burden' was to introduce a technology-neutral quota system, such as the Swedish-Norwegian system of tradable green certificates. Unlike the German feed-in system, this provides incentives to invest in only the *currently* most cost-efficient technologies.

However, a quota system only drives technical change incrementally as the required markets to induce the build-up of new supply chains are not formed until lower-cost technologies have saturated their markets (Bergek and Jacobsson, 2010). In response, it is argued that immature technologies should not be fostered by market formation policies but by R&D policy. For instance, Frondel et al. (2010, 4055) suggest that:

> one should have abstained from strongly subsidizing the market penetration of relatively immature PV technologies. Rather, from an economic perspective, R&D funding should have increased first.

It is, however, only in the controversial linear model of innovation that the innovation process is seen as 'flowing smoothly down a one-way street' (Kline and Rosenberg, 1986, 285) where research leads to development, development to production, and production to marketing, and where R&D is therefore sufficient for driving innovation and cost-reductions. Of course, R&D is required throughout the life-cycle of a technology, but it has to be supplemented by market formation in order to stimulate the formation of a capital goods sector and to induce it to conduct R&D, product development, and all other measures that drive down costs (e.g. standardization). Hence, while in the linear model markets materialize *after* a technology is fully developed, real-life technologies *co-evolve* with markets. The limitation of a technology-neutral system is, therefore, that it contributes too little to 'putting new technologies on the shelf' through providing the time and markets required for fostering new capital goods industries. Indeed, Hoppmann et al. (2013. 1000) conclude that, in the case of Germany, solar cell deployment policies:

> are effective instruments for inducing innovation . . . since many effects such as economies of scale, learning-by-doing and *the build-up of an equipment industry*

would be harder—even impossible—to achieve when using conventional R&D support (emphasis added).

A technology-neutral quota system may, however, be attractive for those whose objective is to support a quite *limited* deployment of mature technologies with minimum short-term *private* costs, as in the Swedish case. Yet, there is a considerable risk that a quota system would generate high, and not low, *consumer* costs. This would occur if there are (i) large differences in cost levels of renewables and (ii) limited potential (in relation to the overall deployment goals) of lower-cost technologies. The certificate price would then be set by the currently highest-cost technology—offshore wind power in the case of Germany—which would generate rents for investors in lower-cost technologies such as onshore wind power (Bergek and Jacobsson, 2010). Indeed, these rents have been argued to be a strong factor against supporting offshore wind power with quotas in Sweden (Söderholm, 2009). It is, therefore, odd that a debate focusing on *consumer* costs should ignore the risks of such rents.

7.5. Meeting the Triple Challenge: Eight Lessons for Policy

In this section, we identify eight lessons from the German experience of meeting the triple challenge of climate change, globalization, and governance in the field of energy.

A first lesson is that it is not beyond our ability to conceive of and implement a long-term policy addressing the climate and globalization challenges. With EEG and supporting regulations, Germany has demonstrated that it is possible, in technical and economic terms, to transform the electricity sector and to decisively reduce carbon emissions. As the diffusion of renewables, in the German context, must first be counted against nuclear phase-out, this reduction lies largely in the future. A continued diffusion at the average level in the period from 2009 to 2013—i.e. 14 TWh per year of renewable power (AGEB, 2014)—would permit a closing down of nuclear production, and a 75 per cent reduction of coal-based generation, by the mid-2030s if consumption does not increase again.[36] The record is also impressive as a response to the globalization challenge. EEG contributed to the creation of an innovative capital goods industry which developed technologies for which diffusion has been rapid (in Germany and abroad), generating both employment and exports. A continuation of *Energiewende* could also be an answer to the governance challenge. This project is supported by the population at large,

[36] Natural gas generation of about 90 TWh (a level reached in 2008) and increased storage facilities can balance fluctuating generation. Flexibility will also be gained from small combined heat-and-power plants and demand-side management coupled to smart grids.

despite dissent at the government level (EMNID, 2013; AEE, 2014). In many industrial countries, including major EU countries, popular support for shifting to renewable energy may be of similar strength, as indicated by surveys (e.g. EMNID, 2014). Arguably, renewable energy policy as practised by Germany (at least until 2011) has the potential to help overcome the problem of Europe's political lethargy and lack of self-confidence.

A second lesson is that, as the energy transition proceeds, the key policy challenges take new forms. In the years ahead, meeting these may well require less long-term investment in technological innovation (including financing the growth of an equipment industry), now that the big breakthroughs in wind and solar power have been made.[37] The biggest problem may rather be political—overcoming the resistance of fossil and nuclear interests against their replacement by other technologies. Incumbents who are reluctant to give up their old lines of business argue for a 'level playing field' for all technologies, be they renewable, fossil, or nuclear, which would be created by phasing out of support for renewable technologies while postponing the full internalization of external costs (Enel, et al., 2013; Buttgereit, 2013). In 2000, EEG feed-in tariffs were justified partly by the need to compensate for insufficient internalization (FME, 2000). At that time, full internalization would have created very high costs to consumers, but would not have produced a substantial supply from renewable technologies as these were still in their infancy. The situation is different today: some renewable technologies are well developed and internalization would make them profitable, immediately or within a very short time (Table 7.1). Policy should, therefore, put more emphasis on the internalization of external costs.

The third lesson is that an effective energy transition policy will still require support for investing in some renewable power technologies (depending in part on the extent of internalization). Such a support should not be designed to achieve short-term cost-efficiency but to meet the long-term goals of *Energiewende*—a technological transformation. This requires policies, such as the EEG, which take into account the specific features of each technology and supplement market formation policies with interventions enabling the

[37] Certainly, many technological problems still need to be solved but the costs should be kept in perspective. In 2012–13, critics of the 'old' EEG stressed the 'cost avalanche' to be expected from adapting the distribution grid, from new storage technologies, and from a surcharge growing beyond all bounds. A few months after the new EEG was passed (July 2014), a new study commissioned by the government showed that the distribution grid will require about €1 billion/year if renewable power is to be doubled by 2032, and €2 billion/year if it is to be tripled—a modest amount compared to roughly €20 billion/year of supposed extra cost of renewable power resulting from feed-in tariffs (Solarthemen, 2014a). A source close to the government now claimed that no storage will be really needed for another ten to twenty years (Agora Energiewende, 2014), and energy minister Gabriel inaugurated a utility-size battery that is supposed to make for a cheaper reserve than a coal plant on standby (Deutsche Welle, 2014). October 2014 saw the first decline ever of the EEG surcharge (for 2015).

creation of innovative supply chains. Such policies will not be efficient in terms of short-term *private* costs but they may be in terms of *social* costs, in which subsidies to coal power and the full value of negative externalities are included.

The fourth lesson is that the build-up of a supply chain requires faith in policy continuity. This confidence is, however, easily damaged, which is illustrated by the case of solar cells. EEG 2008 failed to provide for the event of a sudden surge of new photovoltaic installations. When it happened in 2010, this phenomenon was poorly understood. Critics blamed it on excessively high feed-in tariffs that were adjusted too slowly to rapidly falling module prices, and this was indeed part of the problem. However, there was another aspect to the problem. Germany was then the main PV market in the world and many Chinese local and regional governments generously supported the creation of new PV cell and module enterprises—for export, as there was practically no domestic market in China at the time. As a result, they could compete with very low prices with the purpose of gaining market shares abroad. Indeed, competition between 2009 and 2013 was so intense that solar manufacturers were making losses worldwide (Mints, 2014); some of the biggest German and Chinese solar cell producers went bankrupt (e.g. Suntech in China and Q-Cells in Germany). Chinese competitive behaviour was later found to be in conflict with World Trade Organization rules by US and EU authorities (dumping, state aid), but the German Conservative-Liberal government at the time only saw the German PV industry as being unable to stand up to market pressure and let it shrink to half of its size without taking measures other than to steeply reduce solar feed-in tariffs.[38] This has damaged faith in policy continuity among entrepreneurs, within and beyond the PV industry.

A fifth lesson is that it is difficult to maintain an informed and constructive discourse on *Energiewende* and its main policy instruments when it is attacked by powerful opponents who argue that renewables are 'unaffordable' and a threat to industry. Indeed, this critique spilled over to other countries, raising legitimacy questions over this form of support and the associated technologies more broadly in the EU. In particular, the European Commission's (2013, 2) Green Paper discussed policy for 2030 and argued that central considerations for future policies are 'concerns of households about the affordability of energy and of businesses with respect to competitiveness'. Yet, as was shown

[38] It might have granted emergency credit facilities, initiated WTO lawsuits, etc. In contrast, the Chinese Government responded by creating a large domestic market for PV and became the new world leader. Many insolvent German leaders were bought up by South Korean, Chinese, Middle Eastern, or US firms and prospered again, thus Hanwha Q-Cells. The prices for Chinese modules at that time did not reflect the full social cost; this may impact Table 7.1 in a minor way, although this effect was reduced by Jan. 2014 due to price increases for Chinese modules.

in section 7.4, the social cost of renewables, in contrast to consumer cost, is lower than that of coal power. It is, therefore, of fundamental importance that policy-makers refrain from using inappropriate cost concepts.[39]

The sixth lesson is that it is important to distribute short- and medium-term private and consumer costs—which may be considerable for developing new capital goods industries—across as large as possible a number of electricity consumers in order to reduce the risk of harming the legitimacy of initially costly new technologies. We could learn from the solar cell case where Germany led European development (believing correctly that coordinated international action was not forthcoming at that stage) and accounted for half the EU stock and one-third of the global stock in 2012 (IEA, 2013a).[40] German electricity consumers, therefore, greatly contributed to the development of a new capital goods industry which now delivers socially cost-efficient technology worldwide.

A voluntary cooperation agreement, among some members of the EU, to share the cost of developing a new technology would not only distribute the initial 'burden' of taking a strategic technology to the market on more consumers but may also reduce the overall cost of doing so. A recent study of cooperation under the 2009 RES Directive (Ecofys et al., 2014, 30) gives the example of developing technology-specific cooperation between EU members who 'consider offshore wind as a strategic technology option and cooperation would have the goal of facilitating "learning" at the lowest possible cost'. Cooperation would lower costs for the offshore grid transmission infrastructure as well as enable the use of the most cost-efficient locations for offshore wind plants.

A seventh lesson is that care must be taken to handle the transformation so that it does not harm industrial competitiveness or less privileged groups within the population (IEA, 2013). For large energy-intensive firms facing international competition, there are near-total exemptions of the EEG surcharge. In addition, these firms are exempted from electricity tax and grid fees and buy their electricity directly on the exchange. Their prices vary and are usually confidential.[41] A recent study, however, shows German industrial prices of 4–5 Ct/kWh at the lower end, only a fraction of what non-exempted industry pays (Küchler and Wronski, 2014), and much below the social cost (Table 7.1). Non-exempted firms (mostly SMEs) pay the 'regular' industry tariff

[39] Faced with such critics, it is also important to keep the costs of decarbonizing electricity in perspective. OECD/IEA estimated this cost at 0.3% of GDP for Germany; this compares with 0.09% for the UK or 0.14% for Denmark, and 0.24% for Estonia. At the same time it estimated Germany's environmentally harmful subsidies at 1.9% of GDP (OECD, 2013, 2014).

[40] Global means here IEA PVPS Countries and Major Other Markets—see table 5 in IEA (2013a).

[41] Greater price transparency is a precondition for an intelligent discourse.

which includes the full surcharge (6.2 Ct) and is currently about 15 Ct (statista.de, 2014).

As *Energiewende* proceeded, the exemptions rose—at a faster rate than the surcharge. To reduce the competitive advantage resulting from these (up to 99 per cent) and to secure minimum financing by industry of EEG costs, the European Commission (2014, §3.7.2.) laid down that industrial consumers may be exempted from 85 per cent of the surcharge as paid by households (with yet higher exemptions for very high electro-intensity firms). Although EEG 2014 reduced exemptions to 80 per cent, it is untenable to maintain these industrial subsidies; prices need to approach the social cost for all consumers. There is, however, no point in inducing a 'carbon leakage', i.e. shifting production to countries with higher carbon emissions. It remains a challenge to find a solution to this dilemma; one candidate may be carbon taxes on imports which then permit an increase in the carbon price within the EU.

For the 'energy poor', an electricity price of 29 eurocents/kWh (as in Germany) is punishing. As the consumer price is made up of many cost items (in 2011, the surcharge constituted 14 per cent of the consumer price), there should be ample leeway to reduce the price by taking away the electricity and value added taxes (about 8 eurocents). The problem of very low income customers can also be solved by an upwards adjustment of the energy element in welfare payments.

The eighth and final lesson is that the European Commission should reflect on its stance towards EEG and *Energiewende* as well as rethink its current policy stance. In the past (1998, 2001, 2007), the European Commission made repeated attempts to ban 'German-style' feed-in tariffs (Lauber and Schenner, 2011), always with the support of Eurelectric, the European association of electricity incumbents. In the name of 'affordability' and 'industrial competitiveness', the Commission now aims to slow down the shift to renewables through (i) unambitious targets for 2030 (27 per cent for renewable energy, just 7 per cent more than for 2020), (ii) authorization of support policies for a maximum of ten years, and (iii) technology-neutral bidding processes for the support of technologies as soon as they have a European market share of 1–3 per cent (European Commission, 2013b, §§116–27).

Until 2011, Germany was a strong opponent of such moves, but much that has been effective in the *Energiewende* policy was abandoned by EEG 2014 (Bundesgesetzblatt, 2014). Indeed, Economic Affairs and Energy Minister Gabriel 'accepted' much of the Commission position on bidding systems to replace feed-in tariffs and stressed in his negotiations with the Commission the need to limit new burdens on industry in the form of contributions to the EEG surcharge. Germany's advance compliance with those rules will in turn strengthen the Commission in enforcing its new guidelines.

This may have an irreversible effect, not only on Germany but EU-wide. First, the unambitious target for 2030, coupled with a failure to internalize external costs, means that the Commission limits the diffusion of socially cost-efficient technologies and delays the dismantling of socially cost-inefficient coal technologies (section 7.4). Second, the new rules for support policies limit the opportunity to assist technologies which are currently more expensive—such as offshore wind turbines and wave power equipment—through their formative phases and well into their growth phases.[42] This may threaten the EU's ability to achieve a carbon-neutral electricity system in time, as well as reduce its ability to meet the globalization challenge in the electricity sector, since the EU capital goods sector would have a less vibrant home market. Hence, the European Commission is putting our ability to meet the triple challenge in the electricity sector at risk.

References

AEE Agentur für Erneuerbare Energien (2014). '92 Prozent der Deutschen wollen den Ausbau erneuerbarer Energien', EMNID, 15 Oct., <www.unendlich-viel-energie.de/92-prozent-der-deutschen-wollen-den-ausbau-erneuerbarer-energien>, accessed 29 Oct. 2014.

AGEB Arbeitsgemeinschaft Energiebilanzen (2013). 'Bruttostromerzeugung in Deutschland von 1990 bis 2013 nach Energieträgern', data as of 12 Dec. 2013, accessed 24 Dec. 2013, only updated files available at <www.ag-energiebilanzen.de>.

AGEB (2014). 'Bruttostromerzeugung in Deutschland von 1990 bis 2013 nach Energieträgern', data as of 7 Feb 2014, accessed 2 Apr. 2014, Only updated files available at <www.ag-energiebilanzen.de>.

Agora Energiewende (2014). 'Stromspeicher in der Energiewende', Berlin, <www.agora-energiewende.de/...Energiewende/Agora_Speicherstudie_Web.Pdf>, accessed 20 Oct. 2014.

Alberici, S., et al. (2014). 'Subsidies and Costs of EU Energy', interim report, 10 Oct., Ecofys, commissioned by the European Commission, <ec.europa.eu/energy/.../20141013_subsidies_costs_eu_energy.pdf>, accessed 17 Jan. 2015.

BDEW Bundesverband der Energie-und Wasserwirtschaft (2013). 'Proposal for a Fundamental Reform of the German Renewable Energy Sources Act', Berlin: German Association of Energy and WaterIndustries, <https://www.bdew.de/.../20140114-pi-mueller-a-fundamental-reform-of-the-eeg-is-a-core-responsibility-for-the-new-german-fed>, accessed 17 Jan. 2015.

Becker, P. (2011). *Aufstieg und Krise der deutschen Stromkonzerne*, Bochum: Ponte Press.

[42] It seems unlikely that technology-specific bidding as envisioned in the new state aid guidelines (EC, 2014, 3.3.1.110) will be an adequate replacement for feed-in tariffs.

Bergek, A., and Jacobsson, S. (2003). 'The Emergence of a Growth Industry: A Comparative Analysis of the German, Dutch and Swedish Wind Turbine Industries', in S. Metcalfe and U. Cantner (eds), *Change, Transformation and Development*, Heidelberg and Berlin: Physica-Verlag, 197–228.

Bergek, A., and Jacobsson, S. (2010). 'Are Tradable Green Certificates a Cost-Efficient Policy Driving Technical Change or a Rent-Generating Machine? Lessons from Sweden 2003–2008', *Energy Policy*, 38: 1255–71.

Blankenberg. A., and Dewald, U. (2013). 'Public Policy and Industry Evolution: The Evolution of the Photovoltaic Industry in Germany', paper presented at the DRUID Academy 2013, <druid8.sit.aau.dk/acc_papers/s2kqvx3aax8h6jtk1ltb9f4j258b.pdf>, accessed 17 Jan. 2015.

BMU Bundesministerium für Umwelt, Naturschutz und Reaktorsicherheit (2009). *Neues Denken—Neue Energie: Roadmap Energiepolitik 2020*, <www.algore2008.de/roadmap_energiepolitik_bund_2020.pdf>, accessed 13 Jan. 2014.

BMU (2013). 'Renewable Energy Sources in Figures', July, <www.erneuerbare-energien.de/fileadmin/Daten_EE/Dokumente__PDFs_/ee_in_zahlen_en_bf.pdf>, accessed 13 Jan. 2014.

Bundesgesetzblatt (2014). Part I, no. 33, dated 24 July.

Bundesregierung (2010). 'National Renewable Energy Action Plan', <ec.europa.eu/energy/renewables/action_plan_en.htm>, accessed 30 Oct. 2014.

Buttgereit, R. (2013). 'The Renewables Revolution is Under Way', *Photon International*, Nov.

CDU, CSU, and SPD (2013). 'Deutschlands Zukunft gestalten', Berlin: Koalitionsvertrag, <https://www.cdu.de/sites/default/files/media/.../koalitionsvertrag.pdf>, accessed 15 Dec. 2014.

Del Rio, P., and Cerdá, E. (2014). 'The Policy Implications of Different Interpretations of the Cost-Effectiveness of Renewable Electricity Support', *Energy Policy*, 64: 364–72.

Deutsche Welle (2014). 'Batterie ist günstiger als Kohlekraft', 16 Sept., <www.dw.de/batterie-ist-günstiger-als-kohlekraft/a-17925864>, accessed 24 Sept 2014.

EC (2008). *Commission Staff Working Paper – The Support of Electricity from Renewable Sources*, COM(2008)19, Brussels: European Commission.

EC (2011). *Energy Roadmap 2050: Commission Staff Working Paper, Impact Assessment*, SEC (2011) 1565, <ec.europa.eu/energy/energy2020/roadmap/.../roadmap2050_ia_20120430_en.pdf>, accessed 18 Jan. 2015.

EC (2013). *Green Paper: A Framework for Climate and Energy Policies*, COM (2013) 169 final, Brussels, <ec.europa.eu/energy/.../20130702_green_paper_2030_en.htm>, accessed 18 Jan. 2015.

EC (2013b). 'Paper of the Services of DG Competition Containing Draft Guidelines on Environmental and Energy Aid for 2014–2020', Brussels, <ec.europa.eu/competition/.../2013...aid_guidelines/paper_en.pdf>, accessed 17 Jan. 2015.

EC (2014). 'Communication from the Commission: Guidelines on State aid for environmental protection and energy 2014–2020', *Official Journal of the European Union*, C200/1, 28 June, <eur-lex.europa.eu/legal-content/EN/TXT/?...52014XC0628(01)>, accessed 17 Jan. 2015.

Ecofys, BBH, and TU Wien (2014). *Cooperation under the RES Directive, Case Studies; Joint Support Schemes*, report delivered to the European Commission, DG ENER, <ec.

europa.eu/energy/.../studies/.../2014_cooperation_under_the_res_directive_case_study_joint_support_schemes.pdf>, accessed 17 Jan. 2015.

EEG Erneuerbares Energie Gesetz (Various years). 'Renewable Energy Act', basic amendments in 2004, 2008, 2010, 2011, 2012, and 2014.

EMNID (2013). 'Emnid-Umfrage zur Bürger-Energiewende', conducted in Sept., <www.die-buergerenergiewende.de/wp-content/uploads/2013/10/emnid-umfrage.pdf>, accessed 29 Mar. 2014.

EMNID (2014). 'Survey in G7 Countries', conducted 22–27 May, <www.greenpeace.de/sites/www.greenpeace.de/files/publications/20140603_opinion_poll_g7.pdf>, accessed 29 Oct. 2014.

Enel, Eni, E.ON, et al. (2013). 'Heads of Nine Leading European Energy Companies Propose Concrete Measures to Rebuild Europe's Energy Policy', <www.eon.com/en/media/news/press-releases/2013/9/10/heads-of-nine-leading-european-energy-companies-propose-concrete.html>, accessed 25 Oct. 2013.

E.ON (2011). *E.ON Offshore Wind Energy Factbook*, Dec., <www.eon.com/.../eon.../e/EON_Offshore_Wind_Factbook_en_December_2011.pdf>, accessed 18 Jan. 2015.

Eurostat (2012). 'Elaboration on "Total Gross Electricity Generation"', <http://epp.eurostat.ec.europa.eu/tgm/table.do?tab=table&init=1&language=en&pcode=ten00087&plugin=1>, accessed 17 Sept. 2012.

EWEA (2013). 'The European Offshore Wind Industry – Key Trends and Statistics 2012', <https://www.ewea.org/.../statistics/European_offshore_statistics_2012.pdf>, accessed 18 Jan. 2015.

FME (Federal Ministry of the Environment, Nature Conservation and Nuclear Safety) (2000). *Act on Granting Priority to Renewable Energy Sources*, <www.wind-works.org/cms/fileadmin/user.../GermanEEG2000.pdf>, accessed 18 Jan. 2015.

Fraunhofer ISE (2013). 'Stromgestehungskosten Erneuerbare Energien', Fraunhofer-Institut fur Solare Energiesysteme ISI, Nov., <www.ise.fraunhofer.de/.../studie-stromgestehungskosten-erneuerbare-energien.pdf>, accessed 18 Jan. 2015.

Frondel, M., Ritter, N., Schmidt, C., and Vance, C. (2010). 'Economic Impacts from the Promotion of Renewable Energy Technologies: The German Experience', *Energy Policy*, 38: 4048–56.

German Advisory Council on the Environment (2011). 'Pathways towards a 100% Renewable Electricity System', Oct., <www.umweltrat.de/.../2011_10_Special_Report_Pathways_renewables.pdf>, accessed 18 Jan. 2015.

Grübler, A. (1996). 'Time for a Change: On the Patterns of Diffusion of Innovation', *Daedalus*, 125(3): 19–42.

Hellsmark, H. (2010). 'Unfolding the Formative Phase of Gasified Biomass in the European Union', Ph.D. thesis, Environmental Systems Analysis, Chalmers University of Technology, Sweden, <publications.lib.chalmers.se/records/fulltext/130446.pdf>, accessed 18 Jan. 2015.

Hoppmann, J., Peters, M., Schneider, M., and Hoffman, V. (2013). 'The Two Faces of Market Support: How Deployment Policies Affect Technological Exploration and Exploitation in the Solar Photovoltaic Industry', *Research Policy*, 42: 989–1003.

IEA (International Energy Agency) (2008). *Deploying Renewables*, Paris: IEA.

IEA (2012). *IEA Energy Statistics: Renewables for OECD Europe and Electricity for OECD Europe*, Paris: IEA.

IEA (2013a). *Energy Policies of IEA Countries: Germany*, 2013 Review, Paris: IEA.

IEA (2013b). *IEA PVPS Trends 2013 in Photovoltaics Applications*, Paris: Report IEA-PVPS T1-23:2013.

IWR (Internationales Wirtschaftsforum Erneuerbare Energien) (2014). 'Trend Meinung: Erneuerbare Energien werden subventioniert—Staat zahlt keinen Cent', <www.iwr-institut.de/...energiewende/erneuerbare-energien-werden-subventioniert-staat-zahlt-keinen-cent>, accessed 6 Aug. 2014.

Jacobsson, S., and Bergek, A. (2004). 'Transforming the Energy Sector: The Evolution of Technological Systems in Renewable Energy Technology', *Industrial and Corporate Change*, 13(5): 815–49.

Jacobsson, S., and Karltorp, K. (2013). 'Mechanisms Blocking the Dynamics of the European Offshore Wind Energy Industry: Opportunities for Policy Intervention', *Energy Policy*, 63: 1182–95.

Jacobsson, S., and Lauber, V. (2006). 'The Politics and Policy of Energy System Transformation: Explaining the German Diffusion of Renewable Energy Technology', *Energy Policy*, 34(3): 256–76.

Jahn, D. (1992). 'Energy Policy, Nuclear Power and New Politics in Sweden and Germany', *Environmental Politics*, 1(3): 383–417.

Kline, S., and Rosenberg, N. (1986). 'An Overview of Innovation', in R. Landau and N. Rosenberg (eds), *The Positive Sum Strategy: Harnessing Technology for Economic Growth*, Washington, DC: National Academy Press, 275–305.

Küchler, S., and Meyer, B. (2012). 'Was Strom wirklich kostet', Forum Ökologisch-Soziale Marktwirtschaft (FOS), Berlin, <www.foes.de/pdf/2012-08-Was_Strom_wirklich_kostet_lang.pdf>, accessed Sept. 2013.

Küchler, S., and Wronski, R. (2014). 'Industriestrompreise in Deutschland und den USA', Forum Ökologisch-Soziale Marktwirtschaft, Berlin, <www.foes.de/.../2014-07-FOES-Industriestrompreise-Deutschland-und-USA.pdf>, accessed Nov. 2014.

Laird, F., and Stefes, C. (2009). 'The Diverging Paths of German and US Policies for Renewable Energy: Sources of Difference', *Energy Policy*, 37: 2619–29.

Lauber, V. (2011). 'The European Experience with Renewable Energy Support Schemes and their Adoption: Potential Lessons for Other Countries', *Renewable Energy Law and Policy Review*, 2(2): 121–33.

Lauber, V., and Buschmann, M. (2013). 'Germany: Challenges of a Full Transition to Renewable Energy', in E. Michelina and J. M. Hills (eds), *Renewable Energy Governance*, London: Springer, 295–313.

Lauber, V., and Schenner, E. (2011). 'The Struggle over Support Schemes for Renewable Electricity in the European Union: A Discursive-Institutionalist Analysis', *Environmental Politics*, 20(4): 508–27.

Lovins, A. B. (1976). 'Energy Strategy: The Road Not Taken?' *Foreign Affairs* (Oct.): 65–96.

MacKerron, G. (2011). 'Renewable Energy and Innovation Policies: European Experience', presentation to the international workshop on innovation policies and structural change in a context of growth and crisis Rio de Janeiro, 13–15 Sept., <www.redesist.ie.ufrj.br/p8/semi/sld.../Gordon%20Mackerron.pdf>, accessed Jan. 2015.

Mayer, J., and Burger, B. (2014). 'Kurzstudie zur historischen Entwicklung der EEG Umlage', updated 14 July, Freiburg: Fraunhofer ISE, <www.ise.fraunhofer.de/...-/kurzstudie-zur-historischen-entwicklung-der-eeg-umlage.pdf>, accessed Nov. 2014.

Mints, P. (2014). 'How the Don Quixote Principle Drives Solar', Renewable Energy World, 6 Aug., <www.renewableenergyworld.com/.../2014/.../how-the-don-quixote-principle-drives-solar>, accessed Nov. 2014.

Monopoly Commission (2013). 'Monopoly Commission Publishes Special Report on the Situation of Competition on the Energy Markets', press release, Bonn, 5 Sept.

Morris, C. (2013). 'Impact of Energiewende on the Poor', Renewables International, 25 June, <www.renewablesinternational.net/impact-of-energiewende-on-the-poor/.../68361>, accessed Jan. 2014.

OECD (2013). *Effective Carbon Prices*, Paris: OECD.

OECD (2014). *Germany Keeping the Edge: Competitiveness for Inclusve Growth*, Paris: OECD.

Pahle, M. (2010). 'Germany's Dash for Coal: Exploring Drivers and Factors', *Energy Policy*, 38: 3431–42.

Pahle, M., Fan, L., and Schill, W. P. (2011). 'How Emission Certificate Allocations Distort Fossil Investments: The German Example', *Energy Policy*, 39: 1975–87.

Renewable Energies Agency (2013). 'Renewable Energies: 378.000 Jobs in 2012', <www.unendlich-viel-energie.de/uploads/media/AEE_Erneuerbare_Arbeitsplaet>, accessed Nov. 2013.

Rogol, M. (2011). *Explosive Growth*, Austin, Tex.: Live Oak Book Co.

Sachverständigenrat zur Begutachtung der gesamtwirtschaftlichen Entwicklung (2012). *Jahresgutachten 2012/13*, <www.sachverstaendigenrat-wirtschaft.de/jahresgutachten-2012-2013.html>, accessed Oct. 2013.

Söderholm, P. (2009). *Styrmedel för havsbaserad vindkraft*, ER 2009:09, Eskilstuna: Swedish Energy Agency.

Solarförderverein (2013). <www.sfv.de/lokal/mails/sj/verguetu/htm>, accessed Dec. 2013.

Solarthemen (2014a). 'Zahl der PV-Arbeitsplätze fast halbiert', issue 425 (5 June): 4.

Solarthemen (2014b). 'Verteilnetze: EE-Integration kein Problem', issue 432 (25 Sept.): 11–12.

Spiegel online (2014). 'Energiekonzerne wollen Bad Bank für ihre Atomkraftwerke', 11 May, <www.spiegel.de/spiegel/vorab/energiekonzerne wollen-bad-bank-fuer-atomkraftwerke -a-968694.html>, accessed May 2014.

Statista.de (2014). <http://de.statista.com/statistik/daten/studie/252029/umfrage/industriestrompreise-inkl-stromsteuer-in-deutschland>, accessed Oct. 2014.

Storchmann, K. (2005). 'The Rise and Fall of German Hard Coal Subsidies', *Energy Policy*, 35: 1469–92.

Suurs, R. (2009). *Motors of Sustainable Innovation, towards a Theory on the Dynamics of Technological Innovation Systems*, Utrecht: Innovation Studies Group, Copernicus Institute, Utrecht University.

The Economist (2013). 'European Utilities, How to Lose Half a Trillion Euros. Europe's Electricity Providers Face an Existential Threat', 12 Oct., <http://www.economist.com/news/briefing/21587782-europes-electricity-providers-face-existential-threat-how-lose-half-trillion-euro>, accessed 16 Oct. 2013.

UBA (Umweltbundesamt) (2008). *Economic Valuation of Environmental Damage*, Methodological Convention for Estimates of Environmental Externalities, Dessau-Rosslau, Germany: Federal Environment Agency.

UBA (2012). *Schätzung der Umweltkosten in den Bereichen Energie und Verkehr*, Dessau-Rosslau: UBA, <https://www.umweltbundesamt.de/sites/.../hgp_umweltkosten.pdf>, accessed Jan. 2015.

VDMA (2013). 'VDMA Photovoltaics Equipment: Few Impulses during the First Half-Year 2013', <http://pv.vdma.org/en_GB/article/-/articleview/2337966>, accessed Jan. 2015.

VDMA (2014). 'VDMA Photovoltaik-Produktionsmittel: Dynamisches 1. Halbjahr 2014', <http://www.vdma.org/article/-/articleview/5275811>, accessed Sept. 2014.

Wessendorf, F. (2013). *Photovoltaic Equipment Market Update with a Focus on German and International Machine Makers*, Frankfurt am Main: VDMA.

Wilson, C. (2012). 'Up-Scaling, Formative Phases, and Learning in the Historical Diffusion of Energy Technologies', *Energy Policy*, 50: 81–94.

Abbreviations

AGEB	Arbeitsgemeinschaft Energiebilanzen. Working group on energy balances (formed by several public and private bodies)
BEE	Bundesverband Erneuerbare Energien (German Renewable Energy Federation)
BMU	Bundesministerium für Umwelt (Environment Ministry)
CDU	Christlich Demokratische Union (Conservative Party outside Bavaria)
CSU	Christlich-Soziale Union (Conservative Party in Bavaria)
DG	Directorate general (government department of the European Union)
EC	European Commission
EEG	Erneuerbare Energien Gesetz (Renewable Energy Act)
EnBW	Energie Baden-Württemberg, one of the big four German electricity incumbents
E.ON	One of the two biggest electricity incumbents (not an abbreviation)
FDP	Freie Demokratische Partei (Liberals)
GHG	Greenhouse gases
GWh	Gigawatt hours
IEA	International Energy Agency
IWR	Internationales Wirtschaftsforum Erneuerbare Energien
MP	Member of Parliament (here of *Bundestag*, i.e. lower house)
MWh	Megawatt hour

RWE	Rheinisch Westfälische Elektrizitätswerke (one of the two biggest German electricity incumbents)
SPD	Social Democratic Party of Germany
TWh	Terawatt hours
UBA	Umweltbundesamt (Federal Environment Agency)
VDMA	Verband Deutscher Maschinen- und Anlagenbau (Association of German Machinery and Plant Industries)

8

EU Policy and Governance

Part of the Problem or Part of the Solution?

Iain Begg

8.1. Introduction

The European Union (EU) likes to portray itself as a global leader in confronting the challenges of climate change and, as the largest trading bloc, has a strong interest in the open markets that characterize globalization. However, the EU is a governance entity that has limited policy capacity in its own right and has manifestly struggled to deal with the consequences of the crises confronting the Union after 2008. What 'Brussels' proposes and agrees has to straddle the collective priorities of the Union as a whole and the often divergent preferences of (now) twenty-eight Member States. As a result, there are often unrealistic expectations of what the EU as a policy actor can do, while much of what is attributed to the EU level should, more accurately, be understood as the outcome of bargains struck between influential Member States.

It follows that an assessment of 'EU policy' has to take account of two conceptually distinct elements of policy-making: what the EU level of governance does directly as an agent; and the more diffuse processes through which a policy paradigm or stance, such as austerity (Blyth, 2013), becomes the norm across the Union. The latter include many of the more contentious principles that have come to dominate the policy agenda, such as the pursuit of hard-line austerity policies in apparent defiance of Keynesian economic logic, or the insistence on market-based solutions. In many policy domains, the EU lacks either the formal competence or the budgetary means to make a difference, so that the distinction also influences the scope for alternatives and means that calls for the EU to 'do something' are likely to be addressed to the wrong actor.

This chapter appraises the efforts at the EU level to confront the economic and climate change challenges, asking whether these efforts have been a help or a hindrance. It argues that, while the diagnoses emanating from the EU level have generally been well-founded, too many of the ensuing policy interventions have been either tangential to the core challenges or have lacked the financial or political clout to make a significant difference. The next section of the chapter explores how the EU can make a difference, then section 8.3 presents a brief overview of relevant trends and how they have been interpreted in the diagnoses underlying EU policy. The following section describes the main elements of EU policy and governance. The chapter then discusses the strengths and weaknesses of the policy mix at EU level and what would be needed to achieve better outcomes.

8.2. The Scope for the EU to Make a Difference

In investigating whether EU policies aimed at countering climate change, and responding to the economic challenges of globalization and the effects of the 2007–8 financial crisis are part of the problem or part of the solution, three main analytical questions have to be elucidated. The first, given the nature of the EU as a multi-level governance polity (Hooghe and Marks, 2001), is to establish which forms of policy intervention to examine. A second issue is whether the policy approach is appropriate for the challenges that have to be faced. This, in turn, has two dimensions: whether the policy stance offers credible ways forward; and whether it is suited to all the Member States, or wrongly imposes one-size-fits-all prescriptions.

Third, there is the straightforward question—though generally hard to answer empirically—of whether the outcomes of the policies can be adjudged to help to solve the sorts of problems investigated in earlier chapters of this book. The EU has tried repeatedly to promote policies aimed at bolstering economic performance, responding to the competitive challenges posed by emerging economies and shifting towards a more sustainable model of economic development (Commission, 2014a). It has done so through diverse governance mechanisms, ranging from the direct funding of major research programmes to the loose coordination of national policies under the umbrella of what might be called 'mega-strategies'. Targets have been set, recommendations galore have been issued, and painfully constructed common positions have been adopted for international negotiations on areas such as climate change. Yet, as the contributions of some of the previous chapters in this book demonstrate, measured by basic indicators such as relative growth rates or shares of international trade and investment, the EU continues to slide down the international league tables and its performance in innovation and patenting has been unexceptional.

The Nature of EU Policy

The influence of EU policies comes about in disparate ways and this tends to cloud assessments of what value they have. Compared to the federal level of government in mature federations such as the US, Canada, Germany, or Switzerland, the fiscal resources of the EU are small (capped at just over 1 per cent of gross national income) and they lack flexibility in terms of changing the pattern of spending. Moreover, the EU budget is obliged by the Treaty to be balanced, precluding it from being used as an instrument for Keynesian stabilization policies. There is some loan-financing available, notably from the European Investment Bank, but it is mainly targeted at projects that fulfil a range of viability criteria and, in practice, that is not a source of funding able to make a decisive difference.[1] Proposals for some form of debt mutualization (such as those of Delpla and von Weizsäcker, 2010; or Commission, 2011) as a means of easing the financing costs of the Member States under the most severe fiscal pressure have consistently been blocked by the creditor nations, amid concerns about the potential moral hazard.

Net fiscal transfers from the EU budget nevertheless provide a valuable boost to public investment in the poorer regions of the Union, reaching 3–4 per cent of GDP in the most generously supported regions, but for most regions they are only a fraction of a percentage point of GDP and, because some 90 per cent of the EU budget is funded by payments from national budgets, they necessarily deplete resources in the latter. The EU's Structural and Investment Funds have, latterly, been orientated towards dealing with climate change and increasing competitiveness, and have been refocused for the 2014–20 programming period on a range of thematic priorities. These include investment in countering climate change, achieving better balanced territorial development, innovation, and social cohesion. However, with an overall budget of 0.4 per cent of EU GDP and so many goals, there is an inevitable risk that these Funds will be too thinly spread. Moreover, their track record in achieving the formal treaty objective of reducing regional disparities and transforming the economic development of supported regions is patchy (Bachtler et al., 2014). Around 7 per cent of public funding for research comes from the research line of the EU budget (hitherto, the successive Framework Programmes, now known as Horizon 2020), with the rest emanating from national sources (Expert Group, 2010).

[1] EIB lending was around €72 billion in 2013 and, typically, around 90% of EIB annual loans are made to EU Member States, with the rest going to partner countries, implying that lending to EU countries was about 0.5% of EU GDP—see: <www.eib.org/about/key_figures/index.htm>. All URLs cited in the footnotes to this chapter were accessed Dec. 2014.

Successive reforms of the EU's common agricultural policy (CAP) have sought to give greater emphasis to rural development and sustainability (the so-called 'greening' of the CAP), although the resources allocated through direct payments to farmers were maintained in real terms throughout the 2007–13 budgeting period and will only fall a little in real terms during the 2014–20 period.[2]

In addition to direct public spending, the effects of the EU level can be classified under six other distinct categories of policy intervention or governance (see Box 8.1).

Box 8.1. THE EU AS A POLICY ACTOR

- *Limited role in public spending,* with interventions for specific purposes, such as agriculture and regional development.
- *Rule-making and market-making* are at the heart of European integration, giving rise to the single market and monetary union, but making less headway in the social domain.
- *Cognitive, through developing policy ideas.* The EU can shape the policy choices made by Member States—through its coordinating role—by developing approaches to key policy areas. An illustration is in relation to climate change, where the EU has been a global cheerleader for emissions curbs and the development of renewable energy, and has pushed through policy packages aimed at countering climate change. The EU is also able to act as a repository of policy ideas, enabling Member States to draw on the wide range of practices and experience in others and synthesize the best elements. Examples can be found in at least some of the orientations of the European Employment Strategy.
- *Prescriptive, with the EU level imposing obligations.* Some obligations, for example those relating to competition policy, are treaty-based and overlap with rule-making, although the prescriptive element comes to the fore in implementation, while others derive more from non-binding, 'soft law' processes through what is known as the open method of coordination (Zeitlin, 2009). There is much in the regulatory field (controls on state aid or the need to adhere to standards). In some instances, such as in developing the world's first comprehensive emissions trading system (ETS), the EU has made a significant breakthrough—see section 8.4.
- *Advisory, through recommendations.* This is a side of EU policy that is often criticized because of the ease with which recommendations can be flouted, yet the shake-up of governance since 2010 is intended to boost commitments to respect this sort of policy intervention.
- *Enforcement of rules.* This is achieved through explicit disciplinary mechanisms which may include formal sanctions or the application of conditionality.
- *Representation of the EU in international negotiations,* with the Commission acting as the sole agent for trade policy and playing a leading role in international climate negotiations, starting with the Kyoto Protocol. In this respect, the EU level has an important executive function.

[2] Full details can be found in documents on the website of the Directorate-General for Budget of the European Commission.

A complicating factor is that EU rules and agreements can influence what is undertaken at the domestic level. To some extent, these influences become a policy 'model', sometimes reflecting distinctive national preferences, but often also the result of how successful different ideas are in capturing the attention of decision-makers. For example, in the social policy domain, 'social investment' has been espoused in several EU countries as a novel approach to the welfare state, emphasizing investment in the individual rather than passive benefits for welfare clients. Since the turn of the millennium, this idea has been gaining ground (Hemerijck, 2012; Taylor-Gooby, 2008). When such ideas become mainstream, they can be perceived as EU policies, even though they are more likely to be nation-specific in how they are pursued.

8.3. The Trends Shaping EU Policy Initiatives

Despite being one of the most prosperous and wealthy parts of the global economy, the EU has been beset by pessimism about its economic prospects, a malaise accentuated by its apparent inability to deal with the succession of crises—financial, economic, and fiscal—that have hit it since 2007 and the ensuing rise in unemployment and social problems. Polling data from the Pew Global Attitudes survey show a widespread loss of faith in the EU, with a decline in support for the EU in all the countries covered.[3] Efforts to reform the governance of economic and monetary union have been painful and protracted, with a seemingly endless procession of summit meetings, many labelled as 'the last chance'. Contrary to perceptions, the pace of governance reform has been considerable (Begg, 2013a; Belke, 2013), both at the EU level and in some of the countries worst affected by the crisis (Dervis et al., 2013). However, the legacy of the crisis is likely to be a lasting decline in growth potential, while the progress of the previous decade in boosting employment rates has been reversed (Commission, 2013; Arpaia and Turrini, 2014).

There is a widespread consensus that Europe faces a number of 'grand challenges', notably the ageing of the population, growing competitive pressure from emerging economies seeking to make inroads into Europe's domestic and external markets, climate change, and adapting to various broader societal changes.[4] Globalization and its consequences loom large in these

[3] See <www.pewglobal.org/2013/05/13/the-new-sick-man-of-europe-the-european-union>.

[4] Some of the more significant long-run transformations have been examined in depth in the NEUJOB project, which looks in particular at their likely effect on labour markets—see <www.neujobs.eu>.

debates, as does the need simultaneously to compete and cooperate globally, as emphasized by Lundvall (2009), while also promoting the knowledge economy. In the labour market, youth unemployment has again become a critical problem (Commission, 2013), and Europe is finding it hard to make the progress it so often claims to want in reorientating its economy around knowledge-based activities or a sustainable model consistent with countering climate change. Yet, as will be discussed, EU policies and the processes around them purport to deal with these issues.

These challenges have been analysed in numerous reports produced by experts, high-level groups, think-tanks, and different agencies of government, arguably to well beyond the point of diminishing returns. Among the key elements of the diagnoses (many of which have acquired the status of received wisdom even though, in some cases, the underlying evidence is open to doubt) are that:

- Europe is lagging behind its major competitors, such as the US, in generating leading-edge technologies, and is increasingly vulnerable to the rise up the innovation ladder of large emerging markets countries, such as China, Brazil, and India. A common rhetorical question is 'Why is there no European Google, Intel, Microsoft or Facebook?'
- Education systems in Europe are not delivering the 'right' kind of graduates and not doing enough to stimulate the right forms of tertiary education (though for a critique, see Beblavy et al., 2013).
- Systems of innovation lack capacity (Aho, 2006).
- There is a reluctance to take sufficiently seriously the need for a fundamental energy transformation (Fischer-Kowalski et al., 2012) as the long-run basis for dealing with climate change.

Labour markets in many Member States lack flexibility (Commission 2013) and there is too great a reluctance to modernize welfare states, despite substantial moves in this direction (Dräbing, 2013), in ways that can deal with an ageing population, low employment rates for youths, and an increased incidence of those not in employment, education, and training (NEETs). Austerity policies are widely blamed for prolonging the crisis and for ignoring the lessons of history (Blyth, 2013). The nub of the problem has been succinctly expressed by De Grauwe (2014), who argues that there has been a comprehensive misunderstanding of the nature of the crisis. He notes that the early response was (appropriately) a fiscal stimulus, but that subsequently policy-makers have reverted to the prevailing supply-side orthodoxy, itself a reaction to the policy mistakes of the 1970s and 1980s when a supply shock was incorrectly understood as a demand shock.

8.4. Main EU Policy Directions

Since the launch of the Lisbon strategy in March 2000, EU leaders have agreed a variety of ostensibly wide-ranging initiatives aimed at transforming both the content and modalities of economic and social policy-making in ways that are intended to address the challenges discussed in this book—see Box 8.2. Their focus on structural and social policies was initially motivated by the twin needs to reform the supply-side of the economy to enable the EU to be globally competitive and to modernize social models, and was subsequently broadened to include climate change goals.

The intensity of coordination has varied among strategies and over time, although the trend has been for the EU level to become more prescriptive and

Box 8.2. THE EU'S MAJOR STRATEGIC POLICY INITIATIVES

In rapid succession, EU leaders put in place broad initiatives encompassing:

- The Lisbon strategy, as originally agreed in 2000 and revised in 2005 with a closer focus on growth and jobs. The strategy was fleshed out in a set of Integrated Guidelines which had the purpose of orientating the National Reform Programmes (NRPs) that each Member State was required to draw up and implement as part of the strategy. These covered a wide range of policy areas, such as energy transformations, labour market regulation, the sustainability of pension systems, or innovation.
- The Sustainable Development Strategy (SDS), initially agreed at the 2001 Gothenburg European Council, then revised in 2006. While the SDS emphasized environmental aims, alongside social cohesion and competitiveness, as internal EU objectives, it also reflected the global dimension of sustainability, particularly around the issue of climate change.
- An Energy Policy for Europe, put forward by the spring 2007 European Council, that came about partly as a response to growing alarm about the threats from climate change, but was also motivated by both competitiveness concerns and geo-political worries surrounding the perceived unreliability of the principal suppliers of hydrocarbons to the EU. In parallel, the Emissions Trading System (ETS) has been through three stages of development, as explained in section 8.4 (see also Chapter 6).
- An integrated strategy towards social protection and social inclusion (SPSI) that gathered together previously separate processes. Here too, strategic guidelines were agreed at European level along with various reporting and peer review processes, but Member States were free to formulate and implement their own policy strategies.

These strategic initiatives have in common that the bulk of the policy effort was expected to come from the Member States, with the EU level responsible for establishing policy guidelines, monitoring progress, and issuing recommendations where Member States appeared to be diverging from an agreed path. To a limited extent, the EU level also pursued its own policies, complementing the national ones, mainly through regulatory initiatives. However, the primary governance role of the EU level is that of coordinating national policies.

intrusive. In the original Lisbon strategy, the main ambitions were grandly worded, including the now notorious commitment to make the EU 'the most dynamic, knowledge-intensive economy in the world by 2010', but the means were very limited. There were no explicit lines in the EU budget to advance any of the strategies, with the result that any public investment aimed at Lisbon objectives had to be found elsewhere, notably from the cohesion policy and research areas. For the budgeting period 2014–20, cohesion policy has been rebranded as the EU's investment policy.

Cohesion policy for the period 2014–20 is expected to focus the investment effort of the EU by requiring the Member States and regions to nominate a restricted list of priorities. As explained in European Commission (2014b, xxi), these are: 'innovation and research, the digital agenda, support for SMEs and the low-carbon economy'. A corollary is that relatively less will be spent on infrastructure, but there are additional obligations to earmark funding for dealing with climate change and for human capital enhancement. In principle, these are consistent with the objectives highlighted in this book, but it is important to be clear about the scale of the effort: it is only around a third of a percentage point of gross national income.

In addition, the new Commission headed by Jean-Claude Juncker has announced an apparently substantial investment initiative, the European Fund for Strategic Investment, with a headline total of €315 billion. The idea behind the scheme is for €21 billion of EU money, composed of €16 billion from the EU budget and €5 billion from the European Investment Bank, to be used to guarantee private investment in major projects. It is intended to favour the struggling economies of Southern Europe and to support higher risk projects that promote growth and jobs. The new fund has, however, attracted some sceptical comment in blogs and in the media about whether it is genuinely new money and whether the very high multiplier of fifteen is attainable.

Europe 2020

The adoption in 2010 of the Europe 2020 strategy marked a new departure, both in the core aim for growth to be 'smart, sustainable, inclusive' and in the headline targets and components of the strategy. The overarching aim is to ensure that EU Member States have strategic economic outlooks that are sustainable, both macroeconomically and in relation to social and environmental goals. Two of the five headline targets (spending on R&D and educational attainment) could plausibly be interpreted as consistent with a shift towards a knowledge economy, and thus as an EU response to globalization, insofar as they represent innovative capacity and human capital. A third target, aimed at countering climate change, is to curb energy use and carbon emissions while boosting renewables. The five headline targets included marginally amended versions of targets already set under the Lisbon strategy (notably for the employment rate and for investment in R&D as a proportion of GDP), and newly formulated ones for social, educational, and energy goals. A potentially significant governance innovation was the establishment of seven 'flagship initiatives' covering a range of policy areas—see Box 8.3.

On paper, the components of the Europe 2020 strategy add up to a comprehensive roadmap for the reform of EU economies, predominantly on the

Box 8.3. THE SEVEN FLAGSHIP INITIATIVES (FIs) OF THE EUROPE 2020 STRATEGY

To advance the three elements of 'smart, sustainable, and inclusive', the EU drew up strategic initiatives in which separate sets of policy objectives were set out for the EU level and for the national level.

Three of these FIs are identified with 'smart' goals and, to this extent, are about both raising the underlying growth rate and bolstering competitiveness:

- *Innovation Union* has at its centre the new EU research programme known as Horizon 2020, but also includes a strengthened commitment to improving innovation systems at national level.
- *Youth on the move* concerns the excellence and fairness of tertiary education, but also implies commitments to enhance education at all levels and to facilitate transitions into the labour market, as well as curbing youth unemployment.
- *A digital agenda for Europe* embraces the creation of a single market for digital activities, as well as policies to boost infrastructure and the take-up of digital services.

Two further FIs are orientated towards fostering environmentally sustainable growth and relate to the EU's climate change goals.

- *Resource efficient Europe* is largely self-explanatory, with policies at EU and national level intended to reduce harmful consumption of resources and emissions of carbon, alongside a better functioning market, together with removal of harmful subsidies.
- *An industrial policy for the globalization era* combines a wide range of EU-level goals around competitiveness and resource efficiency with national and EU policies to improve the business environment. It therefore combines competitive and climate change goals.

The remaining FIs are more closely associated with inclusive growth and can be interpreted as being about the social dimension to EU policy and efforts to deal with the challenges posed by unemployment and social exclusion discussed in the introductory chapter of this book.

- *An agenda for new skills and new jobs* is partly pitched around the notion of 'flexicurity'—the neologism derived from the Nordic/Dutch social models which provide for a welfare system (essentially social democratic) that combines flexibility in labour markets with stronger support for the individual.
- *European platform against poverty* refers partly to policies aimed at a more cohesive Europe, but also concerns (especially at national level) the integration of disadvantaged groups and the sustainability of social policy programmes.

supply-side, but also incorporating some macroeconomic aims, notably for fiscal policy. The mix of policies cannot easily be pigeon-holed as a single, coherent ideological or economic model. There is, certainly, a market-promoting thread running through much of Europe 2020, but rather than viewing market and state as being in opposition, the strategy is more about recalibrating the role of the state. This is most evident in some of the social policy elements in the strategy which focus on activating the socially excluded and increasing opportunities.

Overall, there is much in the detail of the flagship initiatives that, if pursued and implemented as set out, has the potential to be transformative in meeting the challenges central to this book.[5] The trouble is the word 'if'. It is already evident that some of the ambitions of the Europe 2020 strategy (notably in relation to the labour market and innovation) will not be realized, although there has been substantial progress towards meeting the energy and emissions targets (Commission, 2014c).

Member States have a considerable degree of discretion in where they place the emphasis in the national reform programmes through which they fulfil their commitments under the Europe 2020 strategy. Some find ways of using the strategy to justify progressive policies and a 'high-road' to economic growth; others use it to justify unpopular reforms, some of which favour liberal interests at the expense of other interest groups. Yet on the evidence of the years since the euro-crisis came to dominate the policy agenda, the strategy has lost visibility, momentum, and (perhaps most tellingly) political attention (Begg, 2013a; Diebold, 2013).

The EU's Emissions Trading System

In 2005, the EU launched an innovative Emissions Trading System through which emissions of CO_2 (and two other greenhouse gases) by industrial producers are regulated. At the same time, a market for these permits was established, enabling new entrants or producers who had used up their quota to buy them and those with a surplus to dispose of them. Box 8.4 briefly describes the system.

In a number of respects, the ETS illustrates the strengths and weaknesses of the EU level in developing policy responses to the challenges examined in this book. It is undoubtedly well-intentioned and imaginative in its goals and design. When it was set up, the ETS led the way globally and there have since been attempts to adopt similar schemes elsewhere, including in Australia, Japan, and in certain US states, demonstrating the EU's 'soft power' to

[5] Similar policy issues are examined in considerable detail in the NEUJOBS project; for an overview of some of the policy implications, see Begg (2014).

Box 8.4. THE EU'S EMISSIONS TRADING SYSTEM (ETS)

- The 'cap and trade' principle behind the ETS is that emitters of greenhouse gases obtain initial allocations of permits which they can then set against actual emissions. If they have a surplus of permits, they can sell them and, if they need more, they can buy them in the carbon market. Attempts by producers to emit more than is allowed by their permits is subject to heavy financial sanctions, as set out in a Directive first adopted by the EU in 2003 and since amended twice. The fine is set at €100 per tonne of emissions of CO_2 or its equivalent in other gases (the standard ETS unit), a level more than four times as high as the carbon price has attained since 2007.
- The first phase of the scheme from 2005 to 2007 was not very ambitious and saw so many permits issued that the effective carbon price fell to zero, rendering the trading element of the scheme ineffective. Permits were also issued free to begin with, with levels of emissions prior to the start of the system used to determine who was given them. Although the second phase (2007–12) saw a marginal cut in permits, the reduction in industrial output associated with the economic crisis again led to the carbon price falling to close to zero. Nevertheless, the market in allowances grew rapidly after 2005 and has reached 70 million per day, with an annual trade of 7.9 billion in 2012, worth €56 billion.
- The ETS is now into its third phase, scheduled to run from 2013 to 2020. (See the Commission website for more details: <http://ec.europa.eu/clima/policies/ets/index_en.htm>.) In this third phase, the most significant developments were the introduction of an EU-wide ceiling on emissions, instead of national ones, and the introduction of auctioning of a proportion of the initial allowances, rather than the free allocations that characterized the first two phases. Initially, 80 per cent of manufacturers were given allowances free of charge, but that share will decline progressively to just 30 per cent by 2020. There is also provision for the cap on emissions to fall at an annual rate of 1.74 per cent from power stations, as a result of which the cumulative cut by 2020 from the 2005 figure will be 21 per cent. As part of an agreement reached in October 2014, the annual reduction will increase to 2.2 per cent from 2021, enabling the EU collectively to achieve a cut in emissions from ETS sectors of 43 per cent compared with 2005.
- The system now covers just under half of EU emissions of greenhouse gases, according to the Commission website; it applies to the power and heat sectors, the main energy-intensive manufacturing sectors, and (at least partly) to civil aviation. It applies across the EU and also to the three EFTA countries (Iceland, Lichtenstein, and Norway), although eight of the ten Central and Eastern European (CEE) countries (Slovakia and Latvia are the exceptions) have made use of a provision enabling them to continue granting free allowances up to 2019 to the power sector.
- Extension of the ETS to civil aviation has proved to be difficult and has resulted in largely free allocations during the current phase of ETS. After fractious negotiations with other parts of the world, the inclusion of civil aviation was limited to flights inside the European Economic Area until 2016, and there are exemptions for operators with low emissions. Many other specific exemptions, such as the curiously named *Small Emitter and Hospital Opt-out Scheme* in the UK, also testify to the complex ETS deals behind the scenes.

- Most of the revenue (88 per cent) from the auction of permits accrues to individual Member States in proportion to their 2005 emissions in areas covered by ETS, but a further 10 per cent goes to the least wealthy Member States. The remaining 2 per cent is a 'Kyoto bonus' which is split between the nine Member States that achieved at least a 20 per cent reduction in GHG emissions since 2005, all of which happen to be CEE countries. It will, though, take until 2027 before all free allowances end, a target set in the legislation. Even in phase II, the revenue was substantial for national governments: for example, the UK public finances benefited by £1.2 billion between 2008 and 2012.
- In a new initiative called NER300, the revenue from 5 per cent of the permits issued between 2013 and 2020 was to be used to co-finance demonstration projects in two areas: carbon capture and sequestration (through which the carbon dioxide from using coal as a primary energy source is prevented from entering the atmosphere) and innovative renewable energy technologies. This was increased to 400 million in the October 2014 agreement and renamed NER400. There are also plans to extend international cooperation in emissions trading.

Sources: European Commission Factsheet on 'The EU Emissions Trading System (EU ETS)', 2013 version; Egenhofer et al. (2012); UK Department of Energy & Climate Change website; European Council (2014).

influence others. Yet it has been hamstrung by lobbying, resulting in concessions both to individual Member States and to business interests which have mobilized effectively.

Moreover, the fact that the carbon price was trading at between €4 and €7 per tonne for the year to June 2014 and has not exceeded €18 since 2009 has meant that there is little financial incentive to curb emissions through energy-saving innovations or to shift to alternative sources of energy. It remains to be seen whether this is a temporary aberration or whether the parameters of the scheme need to be reviewed further to ensure a carbon price high enough to make a substantial difference. These problems are characteristic of the lack of flexibility common in the implementation of many EU policies once the tortuous negotiations needed to arrive at a deal have been concluded (Egenhofer et al., 2012). If nothing else, the evolution of ETS also highlights the long time-scales needed to make what looks like an obvious solution work optimally.

The EU Budget

One of the most vexed of policy areas is the EU budget, condemned nearly a decade ago in the Sapir report (2004) as a 'historical relic', but seemingly incapable of being fundamentally changed. After an extended period of negotiations (and much posturing), the European Council reached an agreement in February 2013 on the budget for the period 2014–20. This agreement on the seven year Multi-annual Financial Framework (MFF) sets ceilings for the amounts to be disbursed annually under different budget headings. Although there are further negotiations on the annual budgets, in practice, the MFF deal is when the main political battles are fought (Begg, 2010). Compared with the previous MFF which ran from 2007 to 2013, there have been some changes in the planned mix of EU spending, although it is important to recall that the budget as a whole amounts to just 1 per cent of EU gross national income (GNI). The budget provides for some spending on policies to counter climate change, notably by requiring that regional policy spending should earmark resources for this purpose, while parts of the 'competitiveness' heading under the budget are intended to boost the EU's global competitive position.

- The allocation for research has been increased by some 40 per cent in real terms as part of a sizeable growth in the share of the budget allocated to the heading 'competitiveness for growth and jobs', but will remain under 0.1 per cent of EU gross national income (GNI) at just over 8 per cent of the budget. Indeed, although it will be just over a third larger than over the 2007–13 MFF, the entire budget for the Competitiveness heading (which includes the research budget) will be just 0.13 per cent of GNI over the seven years of the MFF and was the main area to be cut in last-minute haggling. The research budget was the principal casualty as leaders fought to agree the budget, being pared back by some €10 billion. Given that R&D spending from national budgets, taking a weighted average, is about ten times higher in terms of percentage of GNI, the complementary spending from the EU budget is helpful but hardly game-changing.
- Agricultural support will fall marginally in real terms, but will remain one of the largest budget items at around 30 per cent of the total in the earlier years of the MFF, falling marginally to 28 per cent because of a slight planned reduction in direct payments in real terms.
- Similarly, cohesion policy will stay at around a third of the budget, while there is a further budget line for rural development, which will be larger than the research budget. In practice, as under the 2007–13 MFF, the cohesion budget will be the principal EU source of funding for public investment, with expectations that it will contribute to measures to curb climate change, innovation, and a number of other themes, alongside its original goal of fostering regional convergence.

Although there were extended negotiations between the Council (representing the Member States) and the European Parliament about the budget, the headline total proposed by the Council stayed the same and the main concessions to the Parliament were in regard to rather technical aspects of the budget procedure, with only marginal changes to different elements of expenditure.

Crisis-Related Governance Reform

Much the most significant in terms of policy influence is a range of measures aimed at reshaping the governance of economic and monetary union. They include the 'six-pack' of legislative measures (five regulations and a directive, agreed in 2011—see the website[6] of DG Ecfin of the Commission for more details) designed to strengthen fiscal discipline through fiscal rules at both the national and EU level (including a reformulated Stability and Growth Pact), while also establishing a new macroeconomic imbalances procedure (Belke, 2013). For members of the euro-area, the 'two-pack' (two further regulations) imposes much stronger monitoring on national budgets by requiring draft budgets to be submitted in the autumn of each year for scrutiny by the Commission. Although these measures included more credible mechanisms for the imposition of financial sanctions[7] than had been in place under the original Stability and Growth Pact, their primary intention is to be preventative. The Fiscal Compact, embedded in the intergovernmental treaty on Stability, Coordination and Governance, was rapidly put in place in 2012, reinforcing a German-led push for more extensive monitoring of Member State policies and a greater resort to rule-based policy-making. However, despite all this activism in governance reform, a coherent means of resolving the crisis and countering the rise in unemployment has been conspicuous by its absence. Contrary to popular perceptions, the Fiscal Compact is not the embodiment of austerity but is, in practice, a longer term recasting of fiscal rules and coordination processes (Dullien, 2012) aimed at reducing the burden of debt. Austerity, by contrast, is very prominent in the tough policies forced on Member States receiving bailouts, although even here there are significant differences in approach. Greece (now into a third bailout), Ireland (the first to exit, at the end of 2013), Portugal, and Cyprus had to agree to a 'Memorandum of Understanding' and oversight of their economic policy by the 'Troika' composed of the IMF, the Commission, and the ECB. They were initially subjected to relatively high interest rates on a range of loans (some from hastily created new funds, such as the European Financial Stability Facility—a temporary fund set up at an emergency summit held in May 2010—some from the IMF, some provided

[6] See <http://ec.europa.eu/economy_finance/articles/governance/2012-03-14_six_pack_en.htm>.
[7] Financial sanctions were included in the SGP regulations from the outset in 1997, including what is often referred to as the 'nuclear option' of a fine of up to 0.5% of GDP for a Member State which failed to rein in an excessive deficit, but these were never imposed. The recent reforms offer more graduated sanctions and greater automaticity in their imposition.

bilaterally from other EU countries), but subsequently obtained better terms. In 2012, a permanent rescue fund, the European Stability Mechanism (ESM) was set up, albeit with strong conditionality written into its terms of operation.

At the instigation of the French and German leaders, a Euro Plus Pact was agreed in 2011 under which Member States would undertake to reform labour markets and tax systems, promote innovation, and so on. Its underlying aim was to boost the EU's competitive position in the face of global challenges. Among EU-level decision-makers, the appetite for new initiatives remains strong. The election of François Hollande as French President led to the agreement of a Growth Pact (about which little has been heard since the summer of 2012).

A positive interpretation of all these reforms and initiatives is that they will make European economic governance more coherent and go a considerable way towards resolving the coordination problems that were so apparent in the EU response to the succession of crises of recent years. The more negative view is that they are too one-sided, because they emphasize discipline while neglecting incentives for growth and for investment in new technologies and processes, as espoused in this book. A further criticism is that they are more about form than substance, in that they concentrate on mechanisms of governance rather than objectives such as increased growth and job creation, let alone longer term goals such as dealing with climate change. The tone adopted by leaders such as François Hollande at the European Council meeting held immediately after the May 2014 European Parliament suggests that there is a growing awareness that this needs to change.

Annual Growth Surveys and Ensuing Recommendations

A key element of the governance reforms introduced in response to the sovereign debt crisis that began in late 2009 is the establishment of the European semester, during which the EU level scrutinizes and comments on national policies, including budget proposals, leaving Member States to implement these policies during the second semester of the year. The reason for EU oversight is to assess whether the policy approach in each Member State is well-conceived. For this purpose, the Commission scrutinizes national reform programmes, as it has done since 2005 when this policy methodology was initiated as part of the relaunch of the Lisbon strategy. It then distils its findings into a report known as the *Annual Growth Survey* which is published late in the calendar year. This report provides an overarching assessment of economic developments in the EU.

The main outcome of the European semester is, however, the recommendations to Member States about the direction of their economic policy. They consist of a commentary on the performance of the economy and suggestions for policy reform. Given how concise these recommendations are, their value to Member States is debatable. Brussels insiders assert, however, that bilateral discussions between the Commission and Member States are an important further level of the governance process, though one largely hidden from the view of either public opinion or national political actors. Critics of the approach are often sceptical about how much influence the entire process has on policy decisions or in pushing decision-makers to accept tough but necessary choices.

A systematic examination of the recommendations shows, for example, that they are silent on issues relating to energy efficiency and climate change for several of the richer Member States (Begg, 2013b). For several of the Central and Eastern European countries, by contrast, the advice is to increase environmental taxes. Generally, greater priority is given to macroeconomic stability than to countering the negative social and other consequences of the crisis, although several of the Member States where youth unemployment is high are enjoined to take measures to attenuate it, for instance by addressing the problem of unqualified school-leavers.

The Evolving Role of the European Central Bank

An assessment of the EU role in governance would be incomplete without referring to the actions of the ECB, a subject extensively examined by think-tanks (such as Bruegel or CEPS, both Brussels-based) and on blogs (for example, VOXEU). The Central Bank has long been criticized for concentrating excessively on price stability to the neglect of the real economy (see e.g. Watt, 2008). In common with other financial authorities, the ECB can also be criticized for paying too little attention to financial stability, although with hindsight the ECB is arguably less culpable than the US Federal Reserve or the Bank of England in the extent to which its policies contributed to asset bubbles. In some cases, such as Ireland, national policies were probably more to blame for what became unsustainable property price inflation. Clearly, these two sorts of critiques go in opposite directions: one implying that monetary policy was too tight, the other that it was too loose. The ECB has also been criticized for doing too little to ease the interest rate burden on countries under pressure from financial markets or for not adjusting interest rates to reflect conditions in countries in recession. A broader criticism has been that, unlike other central banks, the ECB resisted quantitative easing (QE) as practised by the Fed until 2015, partly because it struggled to find a formula for QE that is both consistent with the mandate conferred on the ECB by the Treaties and would not court a veto by countries like Germany.

Some of these critiques reflect the flawed nature of the institutional and constitutional set-up of monetary union. The ECB's task is to set monetary policy for the euro-area as a whole, so that by sheer arithmetic, much greater weight has to be given to conditions in Germany than in Greece or Ireland, let alone tiny Cyprus. When the economic trajectories of constituent parts of the euro-area diverge as much as they have recently, setting policy for the average is, manifestly, much more tricky. The ECB is further constrained by treaty prohibitions on purchasing the debt of Member States directly from national treasuries, so that calls for the ECB to emulate the Fed in having a programme of quantitative easing, while perhaps analytically correct, have to recognize the practical difficulty (and implicit threat emanating from Germany) that if the ECB goes too far, it could be found to be in breach of its statutes.

The design of the euro may be flawed and Krugman (2012) may well be correct to speak of the revenge of the optimum currency area, but even if particular policy interventions can be questioned,[8] the ECB is constrained by its formal Treaty mandate. Nevertheless, it has arguably played a pivotal role during the crisis in staving off financial collapse, using a range of 'unconventional' measures. It was quick to pump liquidity into the financial system as early as August 2007 and, despite German (especially Bundesbank) opposition to some of these initiatives, undertook secondary purchases of sovereign bonds from the markets, contributing to curbing the increase in nominal interest rates for Spain and Italy. Its communication strategy also had a pronounced effect in stabilizing bond yields, notably the July 2012 speech in which ECB President Mario Draghi said (in London, ironically) that he would do whatever it takes. Two months later the ECB announced a further measure known as outright monetary transactions, through which it would agree to intervene without limit in the bond markets of countries under attack by financial markets on condition that the country in question adopted wide-ranging structural reforms. Even though none has, at the time of writing, availed itself of this new mechanism (and its compatibility with the treaty was challenged in the courts, only being given a legal seal of approval in June 2015), the mere announcement had a dramatic effect in lowering the spreads between the German bund (the usual benchmark) and Italian or Spanish bonds. The belated resort to QE in 2015 is a further step.

[8] For example, raising interest rates in the summer of 2009 in response to fears about inflation in commodity prices, even though it appeared at the time that a number of euro-area economies were bouncing back from the recession of the previous twelve months, now looks like a substantial policy error. It should also be recalled that the gaping hole in Greek public finances had not then been identified, and that the Spanish banking system was being praised for avoiding the excesses seen elsewhere.

The Next Stages

Yet more governance reform is currently in progress under the broad heading of 'genuine economic and monetary union' (GEMU), a blueprint for which was launched by the Commission late in 2012 (Commission, 2012), following an agreement by the European Council (2012). Despite the many governance innovations already adopted, the rationale for GEMU is that the operation of economic and monetary union was beset by flaws. These included the so-called 'doom-loop' between sovereign debt and banking problems which lay behind the problems that led to the bailouts for Ireland and Cyprus, and which put extreme market pressure on Italy and Spain in 2011–12. Governance failings arose not only during the crisis period, but also beforehand, including weak compliance with existing rules and inadequate cross-border policy coordination. The vagaries of financial market pressures (first too lax, later too harsh) were highlighted. The Juncker et al. (2015) proposals envisage the completion of banking union through common deposit insurance, a collective stabilization capacity (implying elements of fiscal union), closer policy coordination, and, in response to concerns about the transfer of power to unelected EU bodies, attention to democratic legitimacy and accountability.

8.5. Assessment and Conclusions: Do EU Policies Add Value?

It is easy—and frequently tempting—to be dismissive of many of the EU's policies, but it is also essential to reflect on how the EU is expected to make a difference. The EU level of governance does not conform readily to models rooted in fiscal federalism in which the highest tier of government has primary responsibility for stabilization policy or is able to steer public investment to areas of greatest impact. The Lisbon strategy, which exemplifies the coordination role, was widely adjudged to have been a failure and to have had few tangible results, especially in its initial five years up to 2005, and in the ten years since it is hard to pinpoint successes directly attributable to successor strategies. 'Brussels' is also frequently criticized for its regulatory role, with accusations ranging from imposition of stifling red tape (the stereotypical British view) to the imposition of norms and practices that are ill-adapted to the national context (consider the stylized French view that Brussels imposes an unwanted neo-liberal model).

The harsh fact is, nevertheless, that what may look like counterproductive EU policy is the result of incompatible national positions and that it is the institutional weaknesses of EU governance mechanisms, notably the inability of the EU level to coordinate effectively across Member States, rather than wrong policies that are to blame. Nevertheless 'more Brussels' is unlikely to be seen as a call to arms in national capitals. The political economy of reform is such that reform is often easier to secure when there is no alternative, an illustration of the maxim that a 'crisis is too good to waste'. Europe or Europe's leaders often have a good idea of what is needed, but struggle to make it happen. Fiscal union, for example, has not advanced much (with only the most tentative of agreements on a possible new, small pot of money at the December 2013 European Council, with details postponed indefinitely) and no agreement on debt mutualization, with the result that an EU-level capacity for fiscal stabilization is still a distant prospect. Although banking union is making some progress, it is in a form that fails to deal with all the problems identified and, in particular, still leaves individual countries as the ultimate backstop (Darvas and Merler, 2013; Posen and Veron, 2014).

Even in areas where the EU has played a leading role, such as action on climate change, the EU is open to the charge of being insufficiently robust. For example, in the Climate Change White Paper published in 2014, targets for curbing greenhouse gas emissions in the transport sector were abandoned, while the agreement reached at the October 2014 European Council to adopt fresh targets for the longer term is open to the criticism that it avoided the setting of national targets (European Council, 2014).

Competitiveness Challenges

A lesson from the euro-crisis is that the accumulation of competitiveness problems—reflected in the increased relative unit labour costs and slow growth of exports—in Southern Europe is, at least in part, the result of a failure to confront the difficult supply-side reforms (in product as well as labour markets) considered essential to prosper in a currency union (see e.g. Tabellini and Wyplosz, 2006). The Lisbon Council (2013) argues that significant progress is now being made in the worst affected countries, with the greatest changes occurring in Greece, Ireland, Spain, and Portugal. Indeed, Ireland and Portugal have now exited their adjustment programmes. As Dervis et al. (2013) show, both the fiscal adjustment and the supply-side changes have been enormous in Greece, though there is only now a glimmer of hope that it will pay dividends and the advent of the Syriza government seems to have stalled progress.

But in the next phase, the focus will have to shift to sustaining recovery and social cohesion. Greece exemplifies the challenges: having had to make a huge fiscal adjustment, it is now faced with rebuilding a shattered economy. According to Chrysoloras (2013), the two main needs are to avoid overly onerous debt service (perhaps by easing debt terms, rather than an explicit default) and investment to 'expand the country's productive base, increase wealth production, and alleviate the social crisis, until the ongoing structural reforms bear fruit'. According to the Lisbon Council (2013) even Italy, long a laggard in reform, is now outpacing not only France, but also Germany and the other supposedly competitive economies of Northern Europe in the pace of reform. The short-lived Monti government, for example, achieved significant pension reforms and its successors have identified, and are acting to adopt, a range of further measures. In particular, these countries are continuing to make large adjustments of labour costs, belying the widely held view that internal devaluation is not achievable.

The social costs are manifestly large, and the political challenge of carrying electorates remains daunting, as exemplified by the outcome of the February 2013 Italian election and the subsequent fall of the Letta government, but if the Lisbon Council data are given credence, some cautious optimism may be warranted. Yet it is far from obvious that EU policies are either the source of the improvement or that they provide a platform for a longer term reinvigoration of growth.

The EU as an Agent of Policy

The rhetorical question this chapter tries to answer is whether EU policies and governance are a help or a hindrance. Unsurprisingly, a simple 'yes' or 'no' answer cannot be offered, not least because the range of domains in which the EU potentially affects outcomes is so great and, as the taxonomy in Box 8.1 shows, EU influence manifests itself in disparate ways. Nor is there an easy answer to the question of what is an EU policy as opposed to a policy that emerges mysteriously from EU rules, processes, and interactions among Member States.

Being a member of the EU imposes constraints on national autonomy and policy sovereignty, the more so for participants in the euro. This is the inevitable consequence of the core *rule-making* role of the EU. Hence it is unreasonable to argue that, for instance, the ECB should tailor monetary policy to the problems in selected Member States. Being part of a monetary union implies costs as well as benefits, and constraints as well as opportunities or forms of support. A clear distinction therefore has to be made between, on the one hand, whether the objective EU policy was trying to pursue was appropriate for the EU (or the euro-area) as a whole, rather than for individual Member States, and on the other, policy errors—whether by the ECB or others.

However, the limited size of the EU budget means that it generally lacks the capacity to make game-changing public investments. Instead, it shapes responses to challenges much more by establishing and enforcing rules governing economic activity and by coordinating national policies. The national level is then responsible for the bulk of the implementation and, as numerous episodes have shown, is often prone to delinquency in this regard.

The *cognitive* role of the EU inevitably elicits divergent views. To the extent that there is a core economic policy model, it tends to be market friendly and, especially in response to crisis since 2008, has been orientated towards rules for governing macroeconomic policy-making.

- The EU has, on the whole, been effective in identifying the challenges that European countries have to confront, and has been able to provide extensive analyses and diagnoses relevant to policy responses. Although the robustness of such analyses is varied, the EU can nevertheless be adjudged to have a positive *diagnostic function*.
- As a source of *policy innovation*, the EU tends to have a more positive influence. Through governance mechanisms such as peer review, benchmarking, and exchange of information, there is at least scope to diffuse ideas and practices. It is a process that often works more by osmosis than by transparent channels, but in some social policy areas, such as the development of 'flexicurity' and the turn towards social investment, the shifts in national policy thinking can be sizeable.

The *prescriptive* role of the EU has been prominent in developing longer term policy orientations which, ostensibly, focus on the major challenges of dealing with climate change, an ageing population, and underpinning competitiveness.

- At a *strategic level* the EU has had decidedly mixed results. The propensity to design grand strategies which deliver little has to be recognized as a weakness, although there are various reasons for this outcome. Some strategies fall victim to the trap of being 'Brussels talking to Brussels' (Pisani-Ferry, 2006) and have little resonance in national capitals. Member States dutifully prepare their reform plans and report on their implementation, but at arm's length from actual policy-making, and there is an inevitable risk of reporting fatigue or cynicism (resorting to 'copy and paste' in completing the reports they regularly send to Brussels, for example).
- It has had less success in generating policy momentum, partly because of the friction between levels of government and national ambivalence about accepting instructions or prescriptions from 'Brussels', with the result that the translation of diagnosis into action fares less well. A good example is in the area of energy policy where national concerns about security of energy supply or adverse effects on competitiveness have tended to override decisive action on abating carbon emissions.

In its *advisory* role, the EU also has a mixed record. Recommendations to Member States are often worthy, but of limited salience in national policy debates. They are also prone to follow a certain orthodoxy in relation to macroeconomic policies and approaches to labour market policy. A key explanation is that the EU is unusual as a policy-maker because its impact on the challenges facing Europe is less direct than is the case for Member State governments or for the central or federal tier of government in other settings, so that it is circumscribed in its scope to provide *policy leadership*.

In some of the reforms introduced since 2010, the EU level has acquired a more extensive *enforcement* role, although it remains to be seen whether these powers will be deployed.

- The *sanctioning role* has been constrained in the past by drawn-out procedures lasting years; for example, fines under the Stability and Growth Pact would only have arisen after several stages of escalation of demands from the EU level for action by a Member State, and to date none have been imposed. Moreover, inter-institutional rivalries can further inhibit action, weakening the impact EU policies can have.
- Conditionality is more contentious. In some policy areas the EU has the power to impose conditions and, by doing so, can exercise a *disciplining role* more effectively. The trouble is that some Member States are able to bypass these disciplinary mechanisms, especially if they can construct alliances among peers who might want a favour to be returned on a subsequent occasion, possibly on an entirely unconnected matter. In addition, Member States frequently succumb to the temptation either to try to shift blame towards Brussels for unpopular policies or to score political points by rejecting the policy messages.

Although an overall assessment is hard to make, much of what the EU does is, arguably, along appropriate lines. In its *representative* role, it has led globally in trying to advance the climate change agenda, but has had mixed results in international negotiations, as analysed in Chapter 6 of this volume. On paper, the aims of the Europe 2020 strategy seem appealing, both in focusing on a sustainable growth model which pays heed to climate change through the commitment to the three aims of a 20 per cent reduction in carbon emissions, a 20 per cent increase in renewables, and a 20 per cent reduction in energy use, and in preparing the EU to meet the challenge from emerging global competitors. However, political and budgetary constraints limit its impact and mean that the solutions offered are neither sufficiently comprehensive nor sufficiently radical to make a major difference. The many governance changes enacted in response to the euro-crisis, likewise, are designed to provide a more stable environment to support growth and to deal with unemployment, but they have tended to relate more to the longer term policy framework rather than providing immediate solutions.

There is also the enduring governance difficulty that EU rules, processes, and targets are often given little credence in Member States. Nor have the regular calls (for example, as part of the Compact for Growth and Jobs launched in 2012) for the European Investment Bank to raise its capital and to lend more had any obvious impact on EU aggregate investment which continues to languish. It may be that the new EFSI launched at the end of 2014 will make a difference, but, as noted above, the immediate reaction has been rather sceptical. As so often, the net contributors to the EU budget are worried that it will be a pretext to increase the size of the budget and can be expected to resist, and there are clear disagreements on the substance of the plan.[9] Hence, the charge that the EU is part of the problem is most probably exaggerated, but a charge of lack of relevance may be more damning. Simply put, too much that the EU attempts seems to be tangential either to the Member States' policy priorities or to the EU's needs.

Acknowledgements

The author is grateful for financial support from the NEUJOBS project, funded under FP7. The project examines the consequences for the EU economy and jobs of long-term transformations of the energy market, skills needs, and societal changes.

References

Aho, E. (2006). *Creating an Innovative Europe,* Expert Group chaired by E. Aho, Luxembourg: OOPEC.

Arpaia, A., and Turrini, A. (2014). 'Policy-Related Uncertainty: At the Root of the Lost Resilience of European Labour Markets' <www.voxeu.org/article/policy-related-uncertainty-root-lost-resilience-eurozone-labour-markets>, accessed Dec. 2014.

Bachtler, J., Begg, I., Charles, D., and Polverari, L. (2014). *Evaluation of the Main Achievements of Cohesion Policy Programmes and Projects over the Longer Term in Fifteen Selected Regions,* report to DG Regio, European Commission, <http://ec.europa.eu/regional_policy/information/evaluations/index_en.cfm#16>, accessed Dec. 2014.

Beblavy, M., Lehouelleur, S. and Maselli, I. (2013). *Useless Degree or Useless Statistics? A Comparison of the Net Present Value of Higher Education by Field of Study in Five European Countries,* NEUJOBS Working Paper D 4.4.2.A, <www.neujobs.eu>, accessed Dec. 2014.

[9] 'Divided EU Leaders Give Nod to €315bn Juncker Plan', Euractiv, 19 Dec. 2014, <www.euractiv.com/sections/eu-priorities-2020/divided-eu-leaders-give-nod-eu315bn-juncker-plan-310977>.

Begg, I. (2010). *Mollifying Everyone, Pleasing No-One: An Assessment of the EU Budget Review*, SIEPS Economic Policy Analysis, 2010:14. Stockholm: Swedish Institute of European Policy Studies.

Begg, I. (2013a). 'Are Better Defined Rules Enough? An Assessment of the Post Crisis Reforms of the Governance of EMU', *Transfer*, 19(1): 49–62.

Begg, I. (2013b). *Socio-Ecological Transition in a Period of Crisis: How Well is the EU Coping?*, NEUJOBS Policy Report,1, <www.neujobs.eu>, accessed Dec. 2014.

Begg, I. (2014). 'Policy Implications', in M. Beblavy, I. Maselli, and M. Veselkova (eds), *The Future of Labour*, Brussels: CEPS, i. 207–41.

Belke, A. (2013). 'Towards a Genuine Economic and Monetary Union: Comments on a Roadmap', *Politics and Governance*, 1: 48–65.

Blyth, M. (2013). *Austerity: The History of a Dangerous Idea*, Oxford: Oxford University Press.

Chrysoloras, N. (2013). *Greece's Successful Death*, ELIAMEP Crisis Observatory Policy Paper, 16, Athens: ELIAMEP.

Darvas, Z., and Merler, S. (2013). *The European Central Bank in the Age of Banking Union*, Bruegel Policy Contribution, 2013/13, Brussels: Bruegel.

De Grauwe, P. (2014). *Yes it's the Economy Stupid: But is it Demand or Supply?*, CEPS Commentary, 24 Jan., Brussels: Centre for European Policy Studies.

Delpla, J., and von Weizsàcker, J. (2010). *The Blue Bond Proposal*, Bruegel Policy Brief, 2010/03, Brussels: Bruegel.

Dervis, K., et al. (2013). 'Greek Myths and Reality', Brookings Global Economy and Development Programme, <http://www.brookings.edu/research/papers/2013/08/06-greece-recovery-dervis>, accessed Dec. 2014.

Diebold, C. (2013). *Europe's Sustainability Strategy: A Casualty of the Euro Crisis or an Ambitious Restart?*, Bertelsmann Stiftung Policy Brief, 2013/06, Gutersloh: Bertelsmann.

Dräbing, V. (2013). *Welfare Transformation and Work and Family Reconciliation: What Role for Social Investment in European Welfare States?*, NEUJOBS Policy Brief D 5.6, <www.neujobs.eu>, accessed Dec. 2014.

Dullien, S. (2012). 'Reinventing Europe: Explaining the Fiscal Compact', European Council on Foreign Relations Blog, <http://ecfr.eu/content/entry/commentary_reinventing_europe_explaining_the_fiscal_compact>, accessed Dec. 2014.

Egenhofer, C., Marcu, A., and Georgiev, A. (2012). *Reviewing the EU ETS Review?*, CEPS Task Force Report, Brussels: Centre for European Policy Studies.

European Commission (2011). *Green Paper on the Feasibility of Introducing Stability Bonds*, COM(2011) 818, 23.11.2011, Brussels: European Commission.

European Commission (2012). *A Blueprint for a Deep and Genuine Economic and Monetary Union Launching a European Debate*, Commission Communication, COM(2012) 777 final, 28.11.2012, Brussels: European Commission.

European Commission (2013). *Labour Market Developments in Europe 2013*, European Economy, 5/201, Brussels: European Commission.

European Commission (2014a). *Taking Stock of the Europe 2020 Strategy for Smart Sustainable and Inclusive Growth*, COM(2014) 130, Brussels: EC, 5 Mar.

European Commission (2014b). *Investment for Jobs and Growth: Promoting Development and Good Governance in EU Regions and Cities*, Sixth Report on Economic, Social and Territorial Cohesion, Luxembourg: Publications Office of the European Union.

European Commission (2014c). *A Policy Framework for Climate and Energy in the Period from 2020 to 2030*, COM(2014) 015 final, Brussels: EC, 22 Jan.

European Council (2012). *Towards a Genuine Economic and Monetary Union*, Interim Report, Brussels: EC, 12 Oct.

European Council (2014). 'European Council (23 and 24 October 2014) Conclusions', EUCO 169/14, Brussels, 24 Oct.

Expert Group (2010). *Interim Evaluation of the Seventh Framework Programme*, Report to DG RTD, Brussels: European Commission.

Fischer-Kowalski, M., Haas, W., Wiedenhofer, D., Weisz, U., Pallua, I., Posssaner, N., Behrens, A., Serrio, G., Alessi, M., and Weis, E. (2012). *Socio-Ecological Transitions: Definitions, Dynamics and Related Global Scenarios*, NEUJOBS State of the Art Report, 6, <http://www.neujobs.eu/publications/socio-ecological-transition-and-employment-implications/socio-ecological-transitions-de>, accessed Dec. 2014.

Hemerijck, A. (2012). *Changing Welfare States*, Oxford: Oxford University Press.

Hooghe, L., and Marks, G. (2001). *Multi-Level Governance and European Integration*, Lanham, Md.: Rowman & Littlefield.

Juncker, J.-C., Tusk, D., Dijsselbloem, J., Draghi, M. and Schulz, M. (2015). Completing Europe's Economic and Monetary Union. Brussels: 22.10.2015, http://ec.europa.eu/priorities/economic-monetary-union/docs/5-presidents-report_en.pdf accessed 20 July 2015.

Krugman, P. R. (2012). 'Revenge of the Optimum Currency Area', in D. Acemoglou, J. Parker, and M. Woodford (eds), *NBER Macroeconomic Annual 2012*, Chicago: Chicago University Press, 439–48.

Lisbon Council (2013). *The 2013 Euro Plus Monitor: From Pain to Gain*, Brussels: Lisbon Council.

Lundvall, B.-A. (2009). 'The External Implications of the Lisbon agenda', in M.-J. Rodrigues (ed.), *Europe, Globalization and the Lisbon Agenda*, Cheltenham: Edward Elgar, 209–22.

Pisani-Ferry, J. (2006). 'Only One Bed for Two Dreams: A Critical Retrospective on the Debate over the Economic Governance of the Euro Area', *Journal of Common Market Studies*, 44: 823–44.

Posen, A., and Veron, N. (2014). 'Europe's Half a Banking Union' *Europe's World, 27*, <http://europesworld.org/2014/06/15/europes-half-a-banking-union/#.VSfHbl9waHs>, accessed Mar. 2015.

Sapir, A., et al. (2004). *An Agenda for a Growing Europe: The Sapir Report*, Oxford: Oxford University Press.

Tabellini, G., and Wyplosz, C. (2006). 'Supply-Side Coordination in the European Union', *Swedish Economic Policy Review*, 13: 101–56.

Taylor-Gooby, P. (2008). 'The New Welfare State Settlement in Europe', *European Societies*, 10: 3–24.

Watt, A. (2008). *The Economic and Financial Crisis in Europe: Addressing the Causes and the Repercussions*, European Economic and Employment Brief, 3-2008, Brussels: European Trade Union Institute.

Zeitlin, J. (2009). 'The Open Method of Coordination and Reform of National Social and Employment Policies: Influences, Mechanisms, Effects', in M. Heidenreich and J. Zeitlin (eds), *Changing European Employment and Welfare Regimes*, Abingdon: Routledge, 214–45.

9

Innovation as Growth Policy

The Challenge for Europe

Mariana Mazzucato and Carlota Perez

9.1. Introduction

The advanced world is facing a crucial moment of transition. The 2008 bubble collapse left behind the polarization of incomes, high unemployment, low growth, and a fearful financial sector that is steering away from funding the real economy and remaining instead in a casino world, thereby harming the prospects of revival. Indeed, the current emphasis on 'fixing finance', while leaving the real economy sick, risks creating the conditions for the next bubble.

In this chapter, we argue that the theories underlying current policies are misguided and that the aim of returning to 'business as usual' is therefore mired in a fundamental misunderstanding. Current problems are structural and date back to decades before the crisis began. In particular, we take issue with prevailing beliefs about private and public investment and about the role of the State in such investment. We also provide a different narrative of the State, in which what is needed is not just counter-cyclical spending, but an investment-driven, 'mission-oriented' (Foray et al., 2012), and courageous State that can not only guide Europe out of the crisis but also steer and direct growth when it returns (Mazzucato, 2013a). As increasing numbers of policymakers are recognizing, dogged subscription to neo-liberal orthodoxy is a dead end: markets alone cannot return us to prosperity. Our work has shown that investment is driven by innovation; specifically by the perception of where new technological opportunities lie (Pavitt, 1984; Perez, 2002). Private investment only kicks in when those opportunities are clear; public investment must be directed towards creating those opportunities across all policy spaces and affecting the entire economy. Furthermore, the State's role as an investor involves taking risks: in other words, adopting the attitude of 'win some, lose some'. Such risks must be rewarded so that taxpayers not only socialize the risks, but also share in the rewards (Lazonick and Mazzucato, 2013).

We maintain that success in the current transition requires bringing innovation to the centre of government thinking and action. Innovation policy must become growth policy and vice versa. In doing so, innovation-growth policy will affect all other policies: financial market reform, labour market policy, and especially taxation (Mazzucato, 2012, 2013b). A clear understanding of the innovation *potential* inherent in the current historical moment will inform the *direction* that such policies take. Naturally, different pathways can be chosen while moving in this direction, but recognizing the role of policy in making that choice provides us with a better understanding of the 'boundaries' within which civil society and other forces can operate (Stirling, 2009).

This chapter focuses on applying our knowledge of the ways in which innovation occurs (clustered and wave-like; collective; uncertain; and cumulative)[1] in order to understand what must be done to generate long-run growth that is both 'smart' and 'inclusive'. It fundamentally seeks both to understand what has gone wrong and to determine how to repair those failings. We will first look at the role that innovation has played as the main driver of economic growth since the start of the Industrial Revolution, using the long-term lens of technological regimes and paradigms (Dosi et al., 1988; Perez, 2002) to characterize the current transition period in its historical context. These insights enable us to understand that the challenge today is not to 'fix' finance while leaving the economy sick, but rather to change the way that the real economy works. This change must include de-financializing the economy and redirecting investments towards productive mission-oriented areas.

We argue that the way to get the real economy to operate in the current context is to employ a policy direction that is smart, inclusive, and green. Green Growth, as will be discussed below, can become the next big technological and market opportunity, stimulating and *leading* private and public investment. This brings us to a discussion of the role that the government plays in ensuring that such opportunities exist, and particularly the importance of being able to invest—while welcoming the underlying risk and uncertainty—along the entire innovation chain, not just in areas characterized by positive externalities (such as research and development (R&D)). We then offer some criteria for specific fiscal or tax policies in order to achieve such a reorientation (making it more profitable for productive investments

[1] This brief definition draws on histories of innovation studies that have emphasized these characteristics, such as Dosi and Nelson (1994), Freeman (1994), and Lazonick (2009).

and less profitable for speculative ones), and for shaping markets and governance structures to implement such policies. Without smart policies, structures, and behaviours at the organizational level, smart (innovation-led) growth is impossible. We look at the effect that such policies have on steering societal challenges and missions (both technological, as well as societal) and promoting more inclusive growth, where the State not only socializes the risks but also the rewards. We argue that such policies are themselves innovation policies and conclude by summarizing eight key criteria that we believe can help growth policy to be guided by long-run value creation.

9.2. History Matters

It is not enough to agree, as many economists do, that innovation is a key driver of economic growth. In order for this assertion to have real meaning and to produce coherent recommendations across policy arenas, the underlying assumptions of the models and frameworks that guide the policies must be connected with a deep understanding of how innovation actually comes about. Here, a major strength of the Schumpeterian evolutionary theoretical framework is the significant attention it has placed in recent decades on understanding the 'process' of innovation (Freeman, 1994).

Contrary to other investment decisions (such as those relating to stocks, bonds, or business expansion), which entail dealing with a calculable economic risk, innovation is a non-probabilistic process with 'odds' of success or failure that cannot be calculated in advance (Knight, 1921). Furthermore, innovation is not a 'random' variable (i.e. independently and identically distributed) but is instead subject to clustering in systems (Schumpeter, 1912; Freeman and Perez, 1988; Lundvall, 1992). Indeed, it tends to be cumulative and path-dependent, with innovation today building on yesterday's innovation (David, 2004; Dosi, 2005). This is very different from the view in new growth theory (Romer, 1994), where innovation is seen as *risky* (as opposed to fundamentally *uncertain*) and R&D can be modelled as a lottery. Path-dependent innovation, which is clustered and uneven (and statistically associated with fat-tailed distributions), cannot be modelled with classical economic models that assume Gaussian distributions.

In this context, it is crucial to understand the deep sense in which Schumpeter saw innovation as the driver of growth in capitalism. While he claimed that bankers are at the heart of capitalism and that credit creation is what brings expansion (Schumpeter, 1912), he was clear that this was as a response to innovation *opportunities*. In Schumpeter's view, it is not the availability of money that leads to innovation, investment, and growth; it is the availability of potential innovations, as investment opportunities, that brings forth the money (Schumpeter, 1939). Without such opportunities, no amount of tax breaks or quantitative easing will expand the real economy.

Indeed, Pavitt's (1984) taxonomy of innovation places technological opportunities (along with the conditions of appropriability and the knowledge base) as a key driver of investment. This is related to a key finding in industry dynamics that entry rates in new sectors are uncorrelated with current rates of profits; it is the perception of *future* profits and growth rates that drives entry (Dosi and Lovallo, 2007). Indeed, this is what Keynes meant by 'animal spirits' (Keynes, 1936). As opposed to short-term casino-type speculation, production investment is driven by a gut instinct about future technological and market opportunities, not the current bottom line. Therefore, the question for policy-makers should not be how to make it easier for businesses to invest, but how to stimulate their courage and desire to do so. As discussed in Mazzucato (2013a), we are not faced with a lion in a cage that needs to be released by taking away the impediments (tax and regulation); rather, we must first turn what is currently a pussycat into a lion. Indeed, Keynes' assertion that businessmen were not lions, tigers, or wolves, but domesticated animals (Keynes, 1938)[2] poses a very different policy agenda, even for Keynesians. While counter-cyclical spending is an important part of the solution, it also involves moving the boom in the right direction, with businesses willing and able to spend in the uncertain technological areas that could be the key drivers of growth in the future.

Great Surges, Technological Paradigms, and Bubbles: Understanding the Context

To understand how to drive public and private investment around mission-led innovations, it is first necessary to understand similar challenges in previous crisis periods. The current emphasis on 'financialization'—that is, too much speculative investment and not enough productive investment—makes speculation and short-termism appear unprecedented (Haldane, 2011). However, this view is theoretically short-sighted because it discards the possibility of learning from history and understanding why, for example, we are now in a period that is more similar to the 1930s than to the 1980s. Applying insights from Perez's (2002) work on technological paradigm shifts, bubbles, and great

[2] 'Businessmen have a different set of delusions from politicians, and need, therefore, different handling. They are, however, much milder than politicians, at the same time allured and terrified by the glare of publicity, easily persuaded to be "patriots", perplexed, bemused, indeed terrified, yet only too anxious to take a cheerful view, vain perhaps but very unsure of themselves, pathetically responsive to a kind word. You could do anything you liked with them, if you would treat them (even the big ones), not as wolves or tigers, but as domestic animals by nature, even though they have been badly brought up and not trained as you would wish . . . ' (John M. Keynes, private letter to Franklin D. Roosevelt, 1 Feb. 1938).

surges of development enables us to grasp the specificities of each historical period of development and how the patterns observed in each of those periods are replicated today.

THE PARALLEL BETWEEN THE 1920s AND THE 1990s–2000s

The crash of 1929 ended the prosperity bubble of the roaring twenties in the USA, just as the NASDAQ crash in 2000 and the collapse of 2007–8 ended the bubble booms of the 1990s and 2000s, respectively. The 1920s had seen the diffusion of the mass-produced automobile, universal electricity, petrochemicals, radio, and the mass production paradigm (Freeman and Perez, 1988). The 1990s and 2000s witnessed the explosion of information and communications technologies (ICT), the spread of flexible production systems (Womack et al., 1990), and the installation of the global internet and its consequences on globalized finance, trade, and production (Boyer, 2000). Both of those real investment booms subsequently turned into 'casino capitalism' (Perez, 2002). In each case, there was a significant shift in the rankings of industries and countries; in each, the bust led to a freezing of 'animal spirits' and revealed the decoupling of the financial world from the real economy (Krippner, 2005; Dore, 2008) as well as causing growing income polarization. On both occasions, a huge innovation potential was installed and ready to flourish but, then as now, 'financialization' and resistance to government action stood in the way.

THE 1930s AND CURRENT TIMES

It is understanding the key role of massive technical change in different booms and collapses that points to a parallel between today's crisis and opportunities and those of the 1930s rather than with the 1980s. The call for unfettered free markets in the 1980s assumed that the cause of the recession was government regulation (Derthick and Quirk, 1985). However, when viewed with an understanding of the pattern of technological revolutions, the stagflation of the 1980s was not caused by excessive government but by a lack of innovation potential. It was the exhaustion of the mass-production trajectories that drove the post-war boom, while the ICT revolution was only just emerging—albeit with great speed and great promise, but without enough weight to drive the whole economy out of decline (Perez, 2002). The pro-market policies enabled the survival of companies with mature technologies. With less regulation, these companies were able to close down facilities, move offs-hore, reduce personnel, pay lower wages and taxes, and regain profitability through cost-cutting (Lazonick and O'Sullivan, 2000). Higher productivity did not occur then due to increasing returns to scale, as would have been expected from the Verdoorn effect (Verdoorn, 1949; Kaldor, 1966), but rather due to the radical elimination of less productive facilities and many jobs, or what Freeman and Soete (1994) called the 'Verdun effect'.[3]

By contrast, the government strategy around microelectronics, and the computer revolution itself, was very active (Block and Keller, 2011) and created increasing opportunities for new entrants. Funding for small businesses through the SBIR programme in the United States (such as Compaq, Intel, and Hewlett Packard), public procurement for high-tech companies, and R&D funding for universities and government labs all laid the groundwork for the key technologies that would lead to the PC revolution and later to the internet revolution.

THE ROLE OF GOVERNMENT IN BOOMS AND BUSTS

Thus, the bubble boom of the 1990s was not the result of less government but rather the explosion of the ICT revolution riding on government-funded technologies ranging from integrated circuits to the internet (Block and Keller, 2011; Lazonick and Tulum, 2011; Mazzucato, 2013a). Equally, the global boom of the 2000s was enabled by governments (national and foreign) providing large amounts of funds and easy credit. Where governments did fail to provide adequate regulation was in the financial sector, with the unfettered proliferation of toxic financial instruments having catastrophic consequences (Perez, 2009). Today, as in the 1930s, both economists and politicians are increasingly concerned with the reluctance of finance to fund the real economy. As Keynes (1931) wrote in 'The Grand Slump of 1930':

> there cannot be a real recovery, in my judgment, until the idea of lenders and the idea of productive borrowers are brought together again... Seldom in modern history has the gap between the two been so wide and so difficult to bridge.

We now turn to the relationship between finance and innovation.

9.3. Innovation and Finance

Today, as in the West in the 1930s, governments across the world are trying to 'rebalance' the economy away from speculative finance and towards productive finance that can nurture growth in the real economy. Figure 9.1 shows the problem, especially in countries like the United Kingdom and the US, where finance outgrew the real economy for years, leading to the crisis (Krippner, 2005; Dore, 2008). Financialization has been caused by two factors. The first

[3] A pun referring to the destruction of the French positions in Verdun by heavy German artillery during the First World War.

is the financial sector mainly lending to itself rather than to the real (production) economy,[4] with the risks being far higher but severely under-priced (Alessandri and Haldane, 2009). The second factor is the corporate sector's concentration on short-term boosts to profitability (through mergers and acquisitions (M&A) activity and divestiture from long-run areas like R&D), and inflating stock prices and stock options (and hence also executive pay) through the increased trend of share buybacks (in the last decade, these have exploded to US$3 trillion for the Fortune 500 companies alone—see Lazonick and Mazzucato, 2013; Lazonick, 2014).

The former factor resulted in bank assets ballooning, but these assets were increasingly fictitious. When asset prices fell and bank equity was wiped out, banks were so highly leveraged that it required only a 3 per cent fall in asset prices for the major bust to occur. Before that happened, however, bank assets and profits expanded relative to the rest of economy, increasing their value-added contributions, as this is measured indirectly by their interest margin (see Figure 9.1 for the UK). The latter has caused company profits to be reinvested in speculative areas and also led to record levels of hoarding,

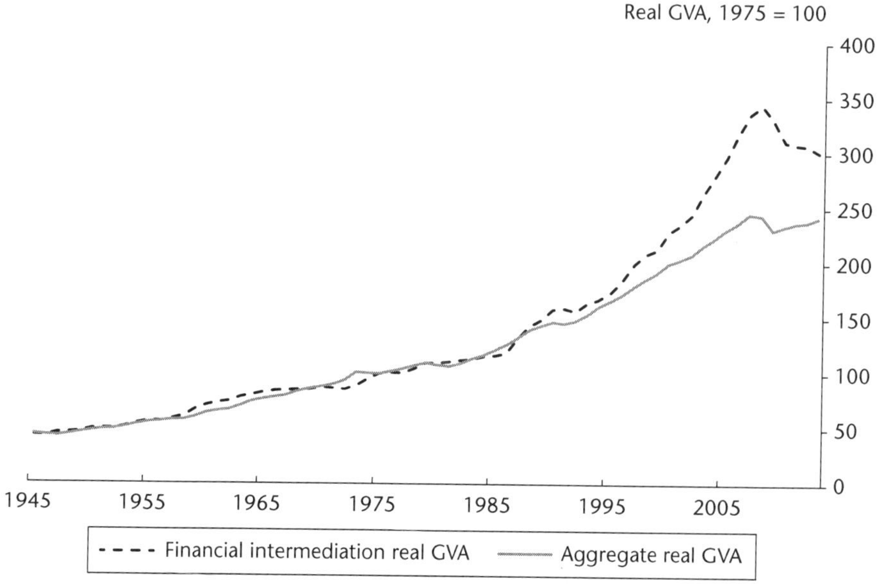

Figure 9.1. Real gross value added (indices 1975 = 100)—UK 1945–2013
Source: Bank of England

[4] By real economy, we mean growth of production in both manufacturing and services that leads to job creation. It is common to distinguish the growth of 'real' value added from value added in finance and agriculture (Mazzucato and Shipman, 2014).

reducing the endogenous creation of new opportunities for future growth: human capital and R&D.

In this context, finance is not playing the role of the 'ephor' of capitalism (Schumpeter, 1934 [1912], 74), and business is not displaying the investment behaviour that Keynes wished for. Instead, the traditional banks and even the venture capitalists have become increasingly risk-averse. Venture capital is focused on the exit from the company, within three years, usually via an initial public offering (IPO) (Lazonick and Tulum, 2011). However, major innovations can take fifteen to twenty years to fully develop, which means that this particular financing model only works for gadgets that ride on existing technology, rather than the big waves of the future. Thus, secular stagnation is a *result* of this financialization, not an excuse for it. Indeed, while a common excuse for buybacks is that they are used when there are no opportunities for investment,[5] if they are indeed reducing the amount that companies are spending on innovation, it can be argued that they are actually *causing* the dismantling of future opportunities, not reacting to the lack of them (Lazonick, 2014). Technological and market opportunities have always been a function of the private and public sectors interacting in dynamic ways. Indeed, a recent MIT report shows that the real problem is the lack of an engaged private sector, with the current absence of large R&D laboratories at private companies, such as those like Xerox Parc and Bell Labs that co-invested alongside the State in the 1960s (Locke and Wellhausen, 2013).

Given the short-termist financial sector and the financialized real economy, policies based on the false assumption of the existence of a shortage of finance are doomed to failure. In order to understand why it is not the quantity of finance but the quality that is important, it is necessary to first look more closely at the relationship between finance and innovation.

Uncertain Cumulative Innovation Requires Patient, Long-Term, Committed Capital

Innovation and finance affect each other. Different kinds of companies and industries, depending on their stage of evolution, require different types of finance, but the finance they actually receive affects their investment patterns and company characteristics (O'Sullivan, 2005; Mazzucato, 2013b). Those companies that want to invest in truly radical new technologies or in sectors in which innovation has a long lead-time (such as pharmaceuticals) require finance that is patient and long-term in nature. Also, because innovation is

[5] Of course, what they mean by 'no opportunities' is no easy-three-year-high-return-opportunities-that-can-yield-as-much-as-financial-gains. It is this distorted market situation that needs to be put right by policy action.

cumulative, finance must always be 'committed'. Across the globe, this long-term funding has often come about through public institutions, whether they are public agencies like DARPA (the Defense Advanced Research Project Agency) in the USA, or through state investment banks like the German KfW or the Chinese Development Bank (Block and Keller, 2012; Mazzucato, 2013a; Mazzucato and Penna, 2014). Indeed, 'mission-oriented' public-sector investments have been fundamental in creating the new opportunities that later drive the private sector to enter new sectors and invest in new technologies (Mowery, 2010).

Understanding the feedback relationship between innovation and the 'type' of finance also helps us understand the relationship between technology and bubbles (Perez, 2002, 2009; O'Sullivan, 2005). Periods during which low-cost finance is widely available but faces few profitable opportunities in the real economy will result in unproductive bubbles, that inflate the value of existing assets (housing, land, wine, paintings, or whatever) and that create casino-type financial instruments to make money with money. Bubbles may still occur when finance is chasing real technological opportunities, but the resulting infrastructures (new railroads, new electrical networks, new internet technology) will continue to fuel growth for decades. What the government provided in the 1990s boom was not easy money but new opportunities through funding the development of new technology (the internet) that constituted a powerful platform for innovation—indeed, a 'general purpose technology' (Lipsey et al., 2005). The high real interest rates prevalent at that time in the US (approximately 7 per cent) were the same as they were during the stagflation of the 1980s, yet this did not deter investors who recognized the above-mentioned opportunities (Figure 9.2a).

During the dot-com bubble, however illusory the promised gains turned out to be, they looked like real opportunities. Therefore, the high cost of credit was no deterrent; at the peak of the bubble, 60 per cent of the IPOs in the US stock markets were in ICT stocks (Figure 9.2b). The fact that real technological opportunities underlay the overinvestment meant that, even though a bursting of the (dot-com) bubble followed, it still left behind a worldwide telecommunications network that facilitated globalization and a social learning process around the web, which enabled many innovations to emerge in social networks and mobile telephony. By contrast, the much lower interest rates (between 2 and 5 per cent) that prevailed in the 2000s (Figure 9.2a) led to the development of new financial instruments for reaping capital gains without having to invest in new (now 'risky') technologies. During that period, easy credit inflated the housing bubbles, while the (much reduced) flow of investment to the stock market involved twice as many finance IPOs as ICT ones.

There is no reason to expect that zero interest rates will bring a better outcome now. Indeed, the results of the recent bouts of worldwide money

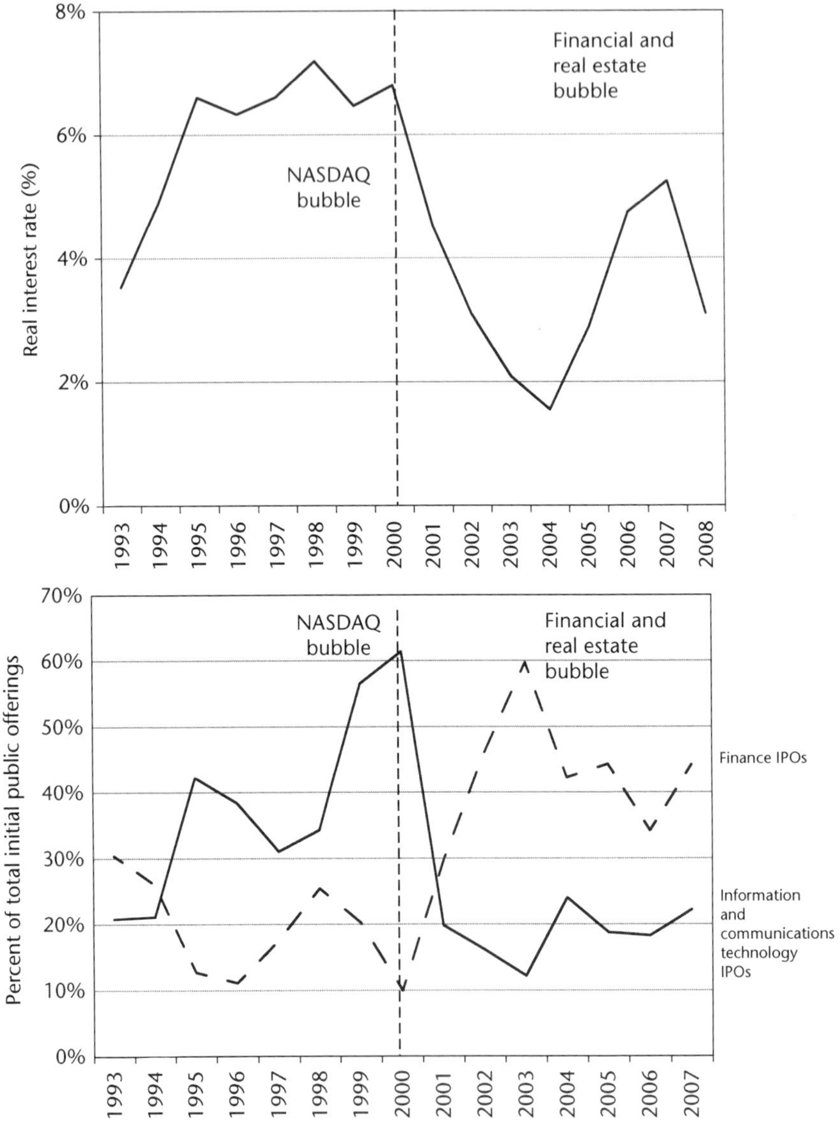

Figure 9.2a and b. Relative real interest rates and % of investment in new technology and finance in 1990s and 2000s bubbles

Source (a): Data from World Development Indicators; our indications

Source (b): Perez 2009 with data from Thomson

printing, euphemistically referred to as quantitative easing, have been meagre and have mainly leaked to the emerging markets (Doukas, 2013; Rajan 2014). This again shows that the revival of real investment does not depend mainly on the amount of finance available or on its low cost, but on the way it is introduced into the economy. Only by directing the creation of money (e.g. by co-financing European Central Bank (ECB) bonds with European Investment Bank (EIB) bonds and guiding the former towards productive investments all over Europe) will the stimulus lead to long-run growth rather than short-run blips.

Supply of Finance versus Demand for Finance

The assumption that there is a finance shortage in both the public and private sides is not only untrue but also leads to inappropriate policies. Today, capital is not scarce (Christensen and van Bever, 2014)—almost $1.2 trillion is being hoarded in the private sector in the US and close to $0.7 trillion in Europe (Zenghelis, 2011)—it is just the wrong type of finance.

A related false assumption, among both mainstream and some heterodox economists, is that there is a 'financing gap'; that is, a credit crunch that requires specific policies to induce the banks to lend. In reality, with regard to important innovation, there is as much a demand-for-finance problem as there is a supply-of-finance problem. By insufficient demand, we mean that there are not enough courageous firms that see clear profit opportunities and are willing to make long-run commitments to high-growth areas that require major innovations (Storey, 2006; NESTA, 2009). It is that sort of demand that can drive the revival of growth and employment. This cannot be achieved by pushing the banks to lend or by making life easier for businesses through various types of tax reductions and tax credits.

The question, then, is not how to provide easy money, but how to bring forth the opportunities, how to mobilize the supply of innovation, how to activate entrepreneurship both public and private. The opportunities will bring forth the money. The emphasis on commercialization assumes that all that is needed is intermediary institutions, and that the quality and quantity of the science base is not a problem. In reality, there has been a massive underinvestment across Europe in key science areas, leading to an obsession with commercialization and knowledge transfer policies without enough knowledge to transfer (Dosi et al., 2006).

It is not sufficient to simply recognize the importance of innovation; it is these accompanying assumptions that matter. The differences in how we understand innovation (path-dependent and cumulative versus a random variable), uncertainty, and the characteristics of investment (driven by the perception of future technological opportunities versus driven by easy and cheap money) will determine the details of innovation policies and, most importantly, our growth policies.

Returning to the 1930s, we see that both Keynes and Roosevelt, coming from completely different positions, not only tried to make investors pull their money out of the banks and bring it into production, but also saw the State as the only entity capable of mobilizing the economy. They argued that this should be done not by giving money to the banks, but by 'pulling' investment and innovation, either through government investment and procurement and/or by putting money into the hands of those with a high propensity to consume. Following Keynes and Roosevelt, such dynamic sources of reliable demand would create the necessary inducement for idle money to invest in production and revive the economy. Innovation was not in their vocabulary, nor in their explicit thoughts; yet the nature of the innovation potential that had been installed during the 1910s and 1920s was inherent to their plans. What made growth actually happen, however, was having a clear *direction* for its deployment.

9.4. The Green Direction

It is important to emphasize the distinction between a *technological revolution*, the *potential* inherent in that revolution, and the *direction* of innovation—and the accompanying investment—in which that potential is deployed (Perez, 2013a, b). Technological revolutions, such as those that brought about the railway age in the nineteenth century, the age of mass production in the early twentieth century, and the information age now, are based on an interrelated set of new technologies, industries, and infrastructural networks that develop on the basis of intensive feedback, providing markets and suppliers for each other (in the way that computers generated markets for micro-chips and the internet generated markets for computers, and so on). In the process, these revolutions also provide a new *potential* to transform and enable innovations in most other industries. We now have smart grids, cars, and soon houses, with the 'internet of things'; nanotech and biotech are dependent on the use of information technologies and can, in turn, help innovate in other industries. Even the globalization of production and trade has been made possible by the internet. However, that potential of ICT as a transformative force across other industries does not have the same self-propelled nature as ICT itself. Its possibilities are disparate and often unconnected; it can be used and shaped in different ways and profitability depends on relative costs and especially dynamic demand. Only a common direction can contribute such synergies. Thus, in contrast with the revolutionary industries, the *direction* chosen for using the new potential across the economy is a matter of socio-political choice.[6] The direction provides convergent trajectories for multiple and disparate industries to innovate, while generating common synergies (suppliers, skills, equipment, service and distribution networks, demand patterns, etc.) that provide advantages for all participants (Perez, 2013b). Suburbanization and the Cold War played that role in the post-war boom in the West; today, Green Growth can serve as a powerful global direction for deploying the potential of the information revolution. If widely understood, such a socio-political choice can reorient innovation across the entire economy, providing a variety of energy sources, a major increase in the productivity of natural resources, and new sustainable lifestyles and patterns of production (Perez, 2013b). It would also enable full global development, given that, if the millions of new consumers entering the market in the so-called emerging countries are to be joined in striving for the American way of life by those of the other developing countries in Africa, Latin America, and the Middle East, they will soon reach planetary limits (Rockström et al., 2009). Only by changing the aspirational lifestyles in a green direction will widespread sustainable development become viable across the whole world.

Innovation Potential, Direction, and Deployment

The effect of a green direction would be the equivalent of the effect that suburbanization and the energy-intensive and materials-intensive lifestyles had on the potential of the mass production revolution after the war. Such successful deployment periods result from purposeful strategic policies. When Keynes said it was fine to simply 'dig ditches and fill them up again', he was desperately trying to put income in the hands of consumers: enabling mass consumption to entice mass production. The specific conditions of his time allowed him to ignore the great importance of the 'direction' of stimulus spending. Yet, the Second World War inadvertently provided this direction by creating massive demand for military innovation and for deploying the mass production of equipment. Having learned that lesson, reluctant businesses (and many equally unwilling politicians) finally accepted State involvement in the economy, at least for the next thirty years or so. Bretton Woods and the Welfare State stabilized international trade and established a reliable expectation of increased disposable income across the national population for the spread of the successive innovations that made possible the American way of life.

[6] The range of political possibilities using the same technological potential can be gauged by the fact that Hitler, Stalin, and the Western democracies established profoundly dissimilar socio-political systems with very different directions for innovation guiding the same mass production paradigm. And among the latter, there was also significant variety, as between the US and Sweden.

To understand the importance of giving a direction to the potential, we must delve once more into the neo-Schumpeterian understanding of innovation. Products are not developed in isolation but as part of technology systems (Freeman, 2008) where each innovation creates problems that call forth solutions from suppliers (Rosenberg, 1983), which then spur investments and can lead to entirely new industries. Universal low-cost electricity brought mass use of refrigerators and freezers, which spawned innovations in frozen foods, which created the need for innovations in packaging methods and materials. This clustering of interdependent users and producers and of self-reinforcing capabilities results in synergies and support networks that make further innovations easier and profitable (i.e. they involve less uncertainty) (Lundvall et al., 2002). At the same time, the aims of 'disposability' and 'reducing work in the home' converged to become common guidelines for innovation trajectories across many industries during the post-war boom. Multiple plastic materials were developed so that dishes and bottles did not have to be washed or returned; synthetic fibres were designed to allow non-iron garments, and plastic disposable packaging was made available for processed, refrigerated, and frozen foods. This encouraged the production of special equipment and the training of personnel for shaping, casting, extruding, and otherwise making it easier to innovate using plastics, which increased the market and reduced the cost through economies of scale. This made it even easier to replace natural materials with synthetic ones when making electrical appliances for kitchen and entertainment. However, without the Welfare State and officially recognized labour unions, which made home ownership and instalment buying a possibility for the great majority, the mass demand would not have been there to generate the scale, the synergies, and the high productivity that made it all possible.

Indeed, suburbanization—based on cheap houses built on cheap land, full of electrical appliances, and with a car at the door—was not only stimulated by roads and other means but often publicly guaranteed (as in the early case of the Federal National Mortgage Association ('Fannie Mae') in the US in 1938). The installed technological potential then benefited from cheap energy and from standardized consumption patterns, enabling economies of scale to be exploited. This lifestyle shift was aided by labour unions ensuring salaries kept pace with productivity increases, along with shorter working hours, unemployment insurance, and pensions. This social security framework, strengthened by the personal credit system, facilitated a steady increase in per-household consumption that brought forth successive series of interrelated innovations spreading across individual domestic markets and overflowing to the rest of the world. In Europe, which had a stronger social democratic tradition, even bolder innovations were spurred by government funding and procurement in national health systems, education, and other public services. Depending on the specific context of each advanced country, there was a varying degree of expenditure on the two aspects of the Cold War—the arms race and the provision of social welfare and good living for the majority of people[7]—both of which contributed to innovation and market expansion.

In essence, the huge innovation potential of the mass-production revolution, which had been installed by the 1930s, was only expressed in growth and prosperity when government provided a direction for its deployment, together with the institutions that sustained it. This is why it is so important to move away from a view of the State as simply 'fixing' market failures to one that sees it making and stimulating investments and mission-oriented innovations (Mowery, 2010). When one asks what missions could drive investment today, and what direction might be given to ensure the full deployment of the ICT industries and their capacity to transform all other activities, one is likely to reach answers that are very different to those arising from doomsday scenarios that see a lack of technological potential, and therefore no substantial opportunities for growth (Gordon, 2012), or from the over-optimistic scenarios that rely on technology and markets to do the job (Brynjolfsson and McAfee, 2011) in the current—financialized—conditions and with just a few educational nudges from the State.

What is Green Growth?

Green Growth is more than just low-carbon and renewable energies. From a technological point of view, renewables alone do not constitute a synergistic technology system. There is not enough technological convergence in knowledge, suppliers, engineering, or skills between solar, wind, wave, geothermal, or hydroelectric energy equipment. In order to benefit from all the potential synergies, the environmental challenge must be viewed through a wider lens. Apart from the technologies that enable flexibility and interaction in the sphere of renewable energy, such as batteries, smart grids, and the like, the green direction would have to encompass what can be termed Green Growth. This would include energy and materials conservation; pollution control; reduction of the material content per product; designing for durability; replacing products with services, possession with rental and maintenance, and waste with recycling; promoting the flourishing of the creative economy; making cities more liveable and less polluting; revamping transport systems and the built environment; promoting collaborative and sharing economies;

[7] Although this way of seeing the Welfare State may be controversial, one can understand that competition between the two systems was being waged at the same time in the arms race and in their capacity to improve the lives of the people.

focusing on health (including preventive and personalized medicine); and promoting all forms of education, both in and out of schools. This type of growth implies a redefinition of the optimal production practices and a fundamentally different view of the 'good life', shaping the desires and aspirations of the majority. In other words, Green Growth involves a gradual transformation of the entire economy, reversing the mass production and consumption patterns of the previous revolution and making it cost-effective and profitable to introduce a wide range of innovative changes in production and lifestyles that would increase sustainability and reduce carbon dioxide emissions, while improving the quality of life for all. Information technologies can be as powerful in enabling those new directions as mass production technologies were for the consumerist directions of the past.

A Political Choice for Growth, Convergence, and Synergies

> The good news is we are sitting on a complete revolution of technology that allows us to move in a sustainable direction. That's a matter of choice though. Markets alone won't take us there. We have to decide on planetary scale, we're going for a sustainable, green, inclusive economy.
>
> (Sachs, 2014)

Why can't markets find the green direction on their own? Because there is no ready-made route that will render the multiple possible directions and disparate innovations profitable. The main characteristic of innovation is uncertainty and it is only in the context of a revolution or when a direction has been clearly chosen and made reality through policy action that the risks will diminish and innovations proliferate in that space as they start to create synergies for each other. There are already innumerable experiments in the private sector, not only in alternative energy, batteries, electric cars, and the like but also in radically new, environmentally friendly production methods, such as the circular economy (Ellen MacArthur Foundation, 2012/2013) or 'cradle to cradle' (McDonough and Braungart, 2002), which aim to eliminate effluents and waste; or industrial symbiosis (Chertow, 2000; Lombardi and Laybourn, 2012), in which different industries use each other's waste as inputs. There have also been significant efforts to stimulate policies in the public sector. Examples include the German, Danish, and Swedish policies on renewable energies; the EU directives on recycling; and tax and other policies in various countries that aim to increase consumer purchases of zero-emission cars or installation of solar panels.[8] However, all of these separate attempts are

[8] Indeed, Tesla cars sell so well in Norway (which buys 80% of them) because the Norwegian Government has made it a convenient and profitable purchase: no tax is paid on purchase; no city parking permit is needed; and their drivers can use taxi lanes.

far from sufficient and evolve too slowly to yield the required growth results. Most of them are too uncertain to attract finance and too dispersed to guarantee demand and generate strong synergies. It all needs an intense collective push to really take off and Europe is a large enough market to be able to make a difference if its member countries act in unison. The problem is that there is still a somewhat fundamentalist understanding of the nature of the free market as neutral and unregulated, when in fact markets would be much more dynamic and profitable if the playing field were clearly and intelligently tilted. Businesses were fiercely resistant to the New Deal and government intervention until they experienced the advantages of receiving state support and a clear direction during the Second World War (Hobsbawm, 1994). This pendular movement for and against government intervention is characteristic of great surges. In the heyday of free markets in the nineteenth century, the British Empire, with procurement practices and subsidies, served as the pacesetter for business and trade expansion and innovation in shipping (Broeze, 1982; Geels, 2002).

The complex feedback loops between government policy, business strategies, and consumer preferences are unpredictable, but without clear and stable policies they may take even longer or not happen at all. What the green direction can do today, through a clear, coherent, and stable set of government policies, is to tilt the playing field strongly towards sustainability. This involves governments taking on the high-risk R&D investments needed to unleash truly new systems, and setting up policies so that demand, taxes, regulation, prices, procurement, and all other contextual elements favour attaining profitability by increasing the productivity of resources. On seeing those opportunities, industries and services would tend to shift their strategies and innovation trajectories to apply criteria of conservation, durability, easy recycling, low-carbon, and environmental preservation.

Such convergence in the direction of innovation would create more and more synergies, making it easier to find suppliers and markets, and to grow and to restore employment levels. For the advanced countries, this may imply focusing on two types of activities: those for high-quality or high-complexity demand sectors (both in equipment and consumer goods or services) that cannot be based on low-cost labour, and those for domestic quality of life that cannot be off-shored. The former can aim to serve the high ends of the consumer markets and the infrastructure, equipment, and engineering needs of the growing developing world. Among the latter, one can count on the greening of the built environment, the sharing and the rental economies, the preventive and personalized health care services, other activities related to quality of life, and so on.

After the 1929 crash, it took more than a decade and a major war before governments finally set in place the institutional innovations that led to the wave of investment innovation and expansion and fully unleashed the potential that had been installed between the 1910s and the 1930s. The same force must be applied in today's context, giving a direction for the full deployment of ICT; this is a challenge that requires a re-examination of the prevailing consensus on the minimal role of the State.

9.5. States as Market-Creating 'Investors', Not Market-Fixing 'Spenders'

The current fashion for keeping government out and for trusting markets to bring prosperity is a misreading of recent experience that does not stem from a deep understanding of the drivers of growth and of the central role played by technical change. Worse still, it ignores the role of the State as the driver of the real-growth miracles of the past few decades: Japan, South Korea, Singapore, China, and others in Asia (Amsden, 1989; Wade, 1990), and Germany and Denmark in Europe. Most of all, it misrepresents the entrepreneurial role of the public sector in driving the ICT revolution in the USA (Mazzucato, 2013a). The global boom of the 2000s was fuelled by the off-shoring and outsourcing movement but not by innovation in new production sectors. In most advanced countries, innovation was centred on the synthetic (and toxic) financial instruments that led to the 2007 housing bubble collapses and to the 2008 Lehman Brothers crash (Perez, 2009). Those bubbles were primarily enabled by governments providing easy credit after 9/11 and by the flows of Japanese and Chinese surplus capital to the West in order to fuel demand for their own export booms. Financial markets were able to exploit this new potential by lifting barriers and removing old regulations. However, no amount of market freedom could have made the private sector risk the huge investments that the US Government made in the development and setting up of the internet in the 1970s and 1980s, or nanotechnology in the 1990s (Block and Keller, 2011).

Drawing on the works of Mowery (2010) and others on mission-oriented innovation, Mazzucato (2013a) has argued that areas of the world that have indeed experienced innovation-led growth have done so through active and strategic innovation policies with direct public spending in a wide range of areas across the entire innovation chain. Such investments were driven by broad visions of change, directed at creating new markets, not just fixing them. In the past this included those investments behind 'the man on the moon' mission, and today we see a similar level of direct investments being aimed at the climate change mission, in countries like China or Germany (with its economy-wide *Energieweinde* strategy). The emergence of phone microchips is due to the procurement of the US military and space programmes, which comprised almost the entire early market for the breakthrough technology, driving the price of the initially costly chips down fifty times in only a few years, thus enabling numerous new applications. Cellular communication itself has its foundations in the radiotelephony capabilities developed during the twentieth century prompted by government support, with Northern European public telecommunications operators persuading 'private corporations, such as Ericsson and Nokia... into a hesitant participation' (Berggren & Laestadius, 2003, 94; see also Block and Keller, 2011). The technologies underpinning the internet, which put the 'smart' into smart phones, were developed and funded by the United States Defence Department's Advanced Research Projects Agency in the 1960s and 1970s. GPS was created and deployed in the 1980s–90s by the US military's NAVSTAR satellite programme, while the revolutionary multi-touch interface was first developed by University of Delaware researchers with support from NSF and CIA grants (Breakthrough Institute, 2010). Even the latest SIRI, iPhone 5's personal assistant, was initially developed with DARPA funding. The same applies in such areas as biotechnology, where private venture capital followed massive public investments that amounted to US$31 billion in 2012 alone (Mazzucato, 2013a).

Growth occurs when a significant technological opportunity is located in a favourable context, and when high-risk, capital-intensive investments are made to kick-start the innovation engine. No amount of price signals would have created the internet, just as today a carbon tax or an emissions market would be useful but certainly not sufficient to get clean-tech going. This is not only about the simple fact that Silicon Valley could not have appeared, grown, and prospered in the old Soviet Union or in Sicily, for example. It is also about the conscious provision of a set of investments, incentives, and conditions that will significantly reduce the risks and increase the potential profitability of what is now technologically feasible but highly uncertain, encouraging convergence in order to create more and more synergies. Policy cannot be limited to temporary counter-cyclical spending, reflecting a shallow understanding of Keynesian advice. In *The End of Laissez Faire* (1926), Keynes defined the role of government intervention in very direct terms: 'The important thing for Government is not to do things which individuals are doing already, and to do them a little better or a little worse; but to do those things which at present are not done at all.' Many companies, both large and small, are currently experimenting in the green direction, but they are far from doing it with sufficient scale or boldly and quickly enough to make a significant difference in growth and jobs. In order for this to be done intensively, kicks and pushes are required, not just gentle nudges, and governments must themselves take courageous risks. Generating consensus in identifying and strengthening context-specific paths in that direction (and the narrower specializations within them) are necessary in order to revive each country's economy in the current global context.

Investment for Innovation and Competitiveness in the Eurozone

The failure to recognize the massive investments that must be made has resulted in incorrect diagnoses of weaker countries, and hence the wrong medicine being prescribed. Consider the Eurozone, where the weakest (peripheral) countries are still facing problems around structural unemployment and low growth. The conventional orthodoxy is that the weaker countries (such as Portugal, Italy, and Spain) had spent too much, but actually these countries have had too little (not too much) spending in areas that create new market and technological opportunities. Figure 9.3 shows that the Mediterranean countries and Ireland have spent below the OECD average on gross R&D in proportion to GDP, which is one of the proxies for how much investment is being made in shaping a dynamic future. Similar measures would be investment in human capital formation (including training programmes) and education.

The weakest (peripheral) countries are also characterized by low business spending on R&D (BERD) in relation to GDP. A key policy area, of course, is how to increase this figure, and there is a common belief that it can be done simply with R&D tax credits. Such credits can help, but since spending on innovation is driven primarily by perceived opportunities, not costs, increasing BERD also tends to require direct public research spending in key areas that the private sector is too risk-averse to fund, but is much more willing to follow up on once the technology is ready for innovation.

The current diagnosis that Europe's problems result from too much State intervention is misdirected; rather the problem is too little strategic involvement and (mission-oriented) investment by the State. And the emphasis on 'liberalization' of companies with State participation ignores the history of many companies that fell in terms of competitiveness when privatized—suggesting that the emphasis should be not public versus private funding but how to build companies which reinvest profits back into areas that foster long-run growth—a matter to which we will return. The current pressure on EU countries not to exceed 3 per cent deficit ratios is putting a strain on economies that need to be investing, not cutting. The 'periphery' (Italy, Greece, Spain, Portugal) must spend more on areas such as education, human capital formation, and R&D. Such investments must be guided by intelligent agencies, driven by 'missions', but also facilitated through EU institutions such as the European Investment Bank, working hand-in-hand with the 'money creation' process of the ECB. Rather than directionless quantitative easing, which has mainly increased money in the banks, not in the economy, the ECB bonds could possibly be co-financed with EIB bonds, as argued by Varoufakis et al. (2013). Indeed, public banks, such as the KfW in Germany and the European Investment Bank (EIB), could become the engine

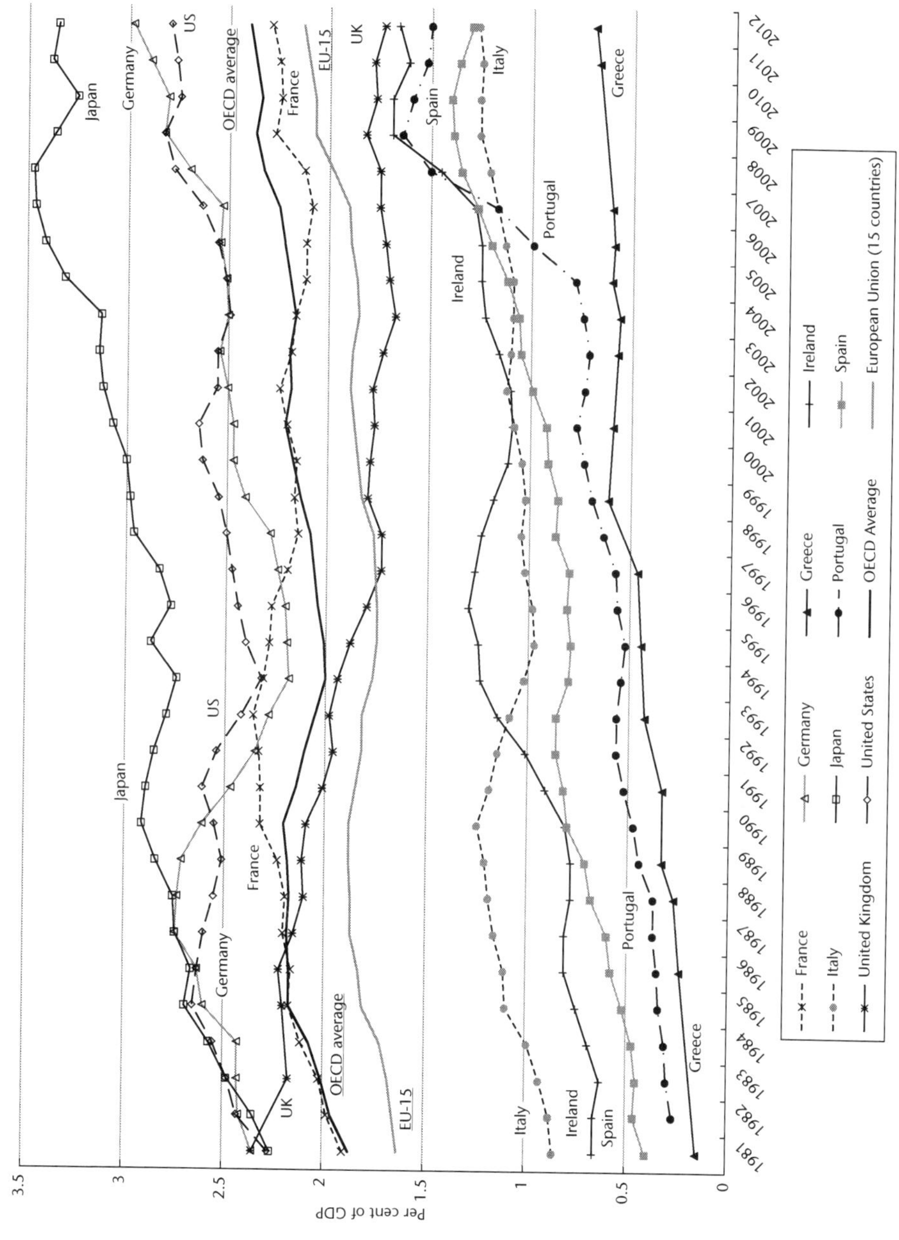

Figure 9.3. Gross R&D spending as % of GDP across OECD countries
Source: OECD

for promoting innovation through long-term patient capital, rather than letting it be impeded by short-term finance. Instead, while the EIB used counter-cyclical lending to kick-start growth from 2008–10, it soon put the brakes on this lending because of concern about credit ratings.

Most of the solutions put forward today are focused on structural reforms and different types of austerity; without strategic public and private investment, however, there will be no sustained growth. As many of the structural reforms also imply cuts to public services and public-sector wages, they are likely to further weaken both demand and the social fabric of countries; this can produce short-term bursts of growth in some countries, but will reduce their ability to escape from the current situation and make them less resilient to future crises.

9.6. Inclusive Growth

Rising inequality, together with unemployment, can be seen as both cause and effect of the current problems and of the recent financial crisis in both developed and developing countries. However, national and transnational ambitions to generate growth that is both smart and inclusive (as proposed in the Europe 2020 programme of the European Commission)[9] are not working. This is not only due to vested interests but also to the lack of a theoretical framework in economics that will allow a clear connection between the type of growth and the level of inequality. While it used to be that economists who studied innovation also studied the distributional impacts (consider the studies of the classical economists on the effect of mechanization on profits and wages, as in Ricardo, 1817), there has been a sharp divide over the last century. Economists interested in innovation have focused primarily on technology, learning, and the dynamics of capabilities, leaving issues of distribution and inequality to be studied by labour economists, mostly from a traditional neoclassical perspective. In turn, governments have also divided innovation policy, labour policy, and economic policy into three distinct and differing worlds (with little or no communication among them).

In recent years, there has been a return to thinking about the relationship between innovation and inequality, through new growth theory, which focuses on 'skill-biased technical change' (SBTC) (Acemoglu, 2002). Apart from methodological concerns with this method (such as the underlying assumption about how labour markets work), a key issue is SBTC's inability to explain the most dramatic increase in inequality between the top 1 per cent

[9] <http://ec.europa.eu/europe2020/europe-2020-in-a-nutshell/priorities/index_en.htm> (accessed Jan. 2015).

(or even more so the top 0.1 per cent) and the rest. In the USA, real incomes for the bottom 90 per cent of earners rose only about 4 per cent from 2000 to 2007, compared to around 94 per cent for the top 0.1 per cent (Piketty and Saez, 2010; Piketty, 2014). Although it is more complex to obtain comparable data across the European Union, the indices that have been developed likewise reveal that the 'top deciles capture an increasing part of the income generated in the economy, while the poorest 10% are losing ground' (cf. European Commission, 2011; Fredriksen, 2012). Is it realistic to explain this polarization in terms of changing levels of skills? We argue that it is not, and that there is a much deeper relationship between the character of innovation and its consequences for the workforce and for income distribution.

Unemployment and the Need for Respecialization

One of the causes of major job loss or displacement is the fact that technical change occurs in clusters corresponding to successive technological revolutions.[10] As the main industries approach maturity and exhaust their ability to stimulate innovation, productivity, or markets, they must shed labour, close down facilities, migrate in search of lower costs and new markets, or disappear entirely (Freeman and Perez, 1988). The modernization option, using the emerging technologies and infrastructures of the next technological revolution, implies radical skill changes, but only appears years later. That was the case with the application of mass-production techniques in the early twentieth century, and more recently with flexible production and the globalization of value chains brought about by the ICT revolution (Van Ark, 2001). These processes are at the root of the changing production landscape, with some regions declining as others flourish, and some skills being devalued (or jobs disappearing) as others gain in value and as new activities find entrepreneurial use for the displaced workers (Brynjolfsson and McAfee, 2011). Understanding the fundamental nature of these changes is essential for designing policies with a reasonable chance of success in the European countries which together with the US have seen the most dramatic upheaval of their previously successful production fabric.

The most extensive job losses in US manufacturing did not occur after the 2007–8 collapse but *during* the easy credit bubble; those were the years of the massive shift of production to Asia. Manufacturing employment in the US

[10] The term respecialization is used here to indicate the process by which companies, countries, or regions find a direction of innovation—and for their production and export profile—that will make them able to compete successfully and sustainably in the globalized economy, in contrast with whatever they were successfully producing in the past.

fluctuated at around 18 million from the mid-1960s to 2000, then dropped by over one-third, from 17 million to 13 million, in the seven years of the easy credit boom (FRED, 2013). Those jobs are not recoverable except through high-tech redesign of a segment of the products involved and/or with the creation of non-off-shorable segments in the economy. The rental economy in the durable goods sector could create hundreds of thousands of jobs in maintenance and distribution, but a return to high growth and high-quality lives will only come about by inducing bold innovation and new technology systems.

Although the job losses may be somewhat less acute in some European countries than in the US, they are of the same nature. In the UK, for instance, there was a greater intensity of job losses in manufacturing during the NASDAQ boom of the 2000s (1 million, or 25 per cent) than during the current recession (nearly half a million) (ONS, 2013). Thus, it is essential to recognize that job losses due to technological change or globalization cannot be recovered with economic 'revival'. Only large-scale innovation can replace these jobs with new ones. And thanks to the MIT project *Production in the Innovation Economy* (Locke and Wellhausen, 2013), it can now be stated—supported by abundant empirical evidence—that without domestic manufacturing activities and capabilities, innovation cannot be turned into production and jobs. It is time to seriously and actively think about respecialization.

New jobs will not be brought about by maintaining austerity policies or by providing a stimulus via the banks. It is only through a major innovation effort, which increases both public- and private-sector commitments to long-run investments in human capital, R&D, and infrastructure, that we can begin to engage in a dialogue in which unemployment is not seen as an unavoidable consequence of technical change. There is nothing 'natural' about 'secular stagnation' (Summers, 2013); it is not an exogenous fate but an endogenous outcome of the lack of investments being made and the lack of an adequate innovation direction being chosen.

Apart from the employment shifts brought by technological revolutions, there is also a strong polarization of income. This is directly linked to the intense financialization of the economy that has historically accompanied the initial decades of diffusion of each successive technological revolution and the resultant major bubbles (Perez, 2002, 2009). The twentieth-century data gathered by Piketty and Saez (2010) on US income distribution among taxpayers, independently of their interpretation, shows how similar the polarization in the recent bubbles is to that of the 'roaring twenties' (see Figure 9.4). Both periods of extreme income inequality are in stark contrast to the resulting reversal in the post-war boom, when suburbanization, the Cold War, and mass production innovation brought jobs and growth, while labour unions, the tax structure, and the Welfare State ensured that the incomes of the

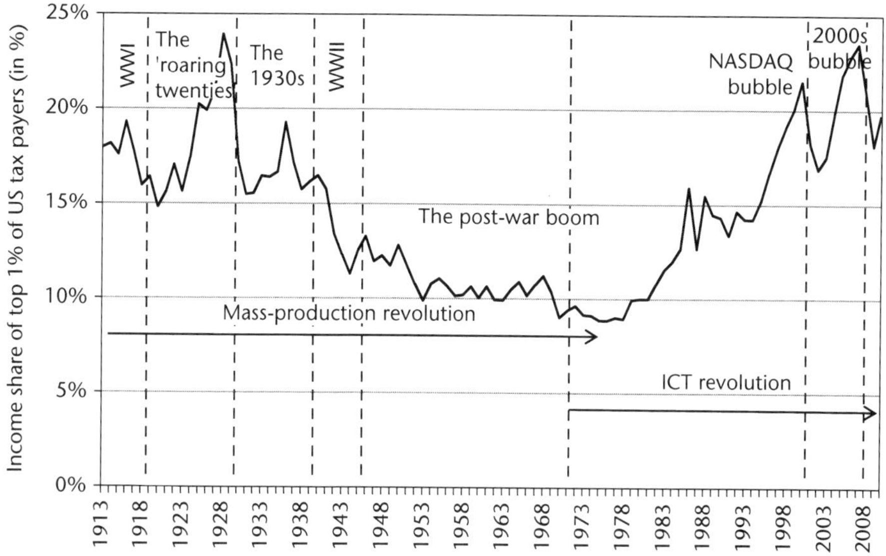

Figure 9.4. Polarization of income during bubbles (US 1913–2010)

Source: Data from Piketty and Saez 2010, Graph and period indications by the authors (based on Perez 2002 and 2009)

majority of citizens increased with productivity. Socio-political choices can make a huge difference.

Lazonick and Mazzucato (2013) have argued that inequality of such an extreme form has to do with the ability of some agents in the economy—specifically the financial ones—to 'extract' value from the innovation process. Hence, the problem of inequality (regardless of whether Piketty's (2014) wealth tax proposal is applied) needs to be faced directly by innovation/industrial policy. It is no use arguing that skill changes are the source of the current inequality. Skills are 'endogenous' to the investment process (i.e. skills are actively created and are not manna from heaven). Therefore, it is necessary to make sure that all investment, and especially public–private partnerships, increase rather than decrease the incentives of private companies to invest in human capital formation, and other long-term areas like research and development.

It is clear that unless industrial policy also asks what kind of innovation ecosystems we want, we risk creating ones that are not so much symbiotic as parasitic—in other words, a system in which a disproportionate share of the rewards from innovation go to a narrow group of actors, who call themselves the great 'value creators' but are in reality merely 'value extractors'. Again, this is a problem inextricably linked to the financialization of the economy.

9.7. Smart Innovation Requires Smart Government

Understanding the problem and the nature of the solution is only a first, albeit indispensable, step. Implementing the necessary directionality also requires effective public actors, and this requires sector and technology-specific expertise to be located in government so that public interventions are not just making it easier (via subsidies) but also raising the stakes and commitment by all and 'making things happen that otherwise would not' (Keynes, 1936). Government must not only fund innovation but also innovate 'within' its own organization and behaviours. One of the most important innovations introduced by government during the Victorian boom in the 1850s was the introduction of exams for entering the civil service (Cohen, 1941). It was the knowledgeable and meritocratic structure of the British Government that allowed such a small country to effectively govern an empire that covered a third of the planet. Before that, posts in the civil service had been bought or granted. Today, many public servants are as competent and highly educated as top managers in private companies (as are many politicians). However, as long as governments are presented as boring bureaucracies that get in the way of creative people and their efforts, they will not nourish existing talent nor be able to attract new talent, which is precisely what is needed in the areas that drive growth.

Another necessary change refers to the ways of measuring and judging performance. Governments, like innovative companies, should accept as normal rather than fear the failures that are inevitable when engaging in uncertain innovations. Those countries that are currently aiming to grow through innovation in particular areas, such as China's and Germany's strategy around 'green innovation', and the focus of countries like Singapore and Brazil in becoming leaders in new markets, face the challenge of attracting sectoral and technological expertise within *top levels* of government, not just in the ministries for innovation and industry. It is crucial for such countries to nurture a trial-and-error approach within the public sector, building innovative organizations. Obama's active post-crisis stimulus programme was (until 2010) heavily influenced by a Nobel Prize laureate scientist, Steve Chu, in the Department of Energy, who was responsible for setting up ARPA-E (Grunwald, 2012). BNDES, Brazil's active and successful state investment bank, which has doubled its lending since the crisis and is directing investments not only in infrastructure but also in key sectors like pharma, biotech, and clean-tech, is run by two (non-neoclassical) innovation economists. By contrast, when the State is consumed with a fear of 'picking winners' and industrial policy is side lined by the Treasury or Ministry of Finance, there is almost a self-fulfilling prophecy that leads to failure, with agencies responsible for industrial policy becoming marginal to the economic strategy and ending up with a patchy innovation policy that is at odds with the macroeconomic growth policy being applied.

9.8. Conclusion: Towards a New European Competitiveness

We have argued that the current recession cannot be treated as an ordinary one, in which the economy can be revived simply through conventional fiscal and monetary policy and, in this case, with financial reform. Europe (and indeed the world) is at a crossroads similar to the 1930s, with major institutional innovations needed to overcome a distorted financialized economy and a society with profoundly polarized incomes. As on that occasion, huge innovation potential must be turned into profitable opportunities, while creating the conditions for a fairer sharing of the benefits across society.

We have also argued that, in order for innovation-led growth to occur, an understanding of how innovation comes about must guide both financial reform and macro growth policy. Smart and inclusive growth must be government-led and innovation-driven, as it has been in the past. It must also go hand-in-hand with changes in corporate governance to provide better incentives for companies to reinvest their profits in long-run value-creation areas like human capital and R&D. Furthermore, government policy must be understood as not only fixing markets but also actively shaping and creating them. This requires providing a clear direction for innovation, investing in it and providing a coherent set of policies (direct and indirect) to tilt the playing field so that it rewards those firms willing and able to invest in the opportunities of the future. Thus smart growth requires a smart—and learning—government and this entails a proactive view of the role of the State (Mazzucato 2015). *Secular stagnation* is in no way inevitable; it is a result of (endogenous) choices that are the opposite of the ones we will outline here.

Our policy criteria can be summarized as follows:

- *Public investment as mission/challenge-oriented.* It is important for innovation policy to be guided more by the need to actively shape and create markets than just to fix market failures or to compensate for imperfect information. This requires the ability to catalyse many sectors to address big 'challenges' (such as tackling climate change, or limits to natural resources, or the ageing crisis). It also means having an overview of the support needed along the entire innovation chain, not just the more obvious 'public good' areas.
- *Direct and indirect investments.* It is unrealistic to think that tax incentives (indirect public spending) alone will work, because business investment is driven by perceptions of opportunities, not just cost. Such opportunities are often created through direct public investments in high-risk areas, into which the private sector is still too risk-averse to enter. As was the

case with the ICT revolution, innovation must be directed along the entire chain, from basic research through applied research to early-stage financing of companies. This also requires admitting that the real crisis in Europe, especially among the weakest nations, is an investment crisis. While structural reforms are necessary, unless the weakest countries greatly increase their investments, labour market reforms and liberalization will have little effect except to redistribute income—regressively.

- *Use the EIB to steer investment-led growth.* Once it has been admitted that a key problem causing skewed competitiveness across Europe is the lack of investment in the periphery, the fiscal compact must be restructured to allow such investments rather than to discourage them. This can be done both nationally and transnationally. The current diagnosis of the problem as one of 'too much debt' rather than 'too little strategic spending' is putting the brake on the EIB, precisely when it could help to kick-start innovation-led growth and provide patient capital in the way that many national development banks do (including the German KfW). The ECB should be ready to step in with its money-creating power to provide bonds that are co-financed with the EIB.
- *Green as the key challenge and redirection.* The 'green transformation' will only occur if such directed challenge/mission-oriented investments can be geared towards green objectives (as defined in section 9.4). If sustainable growth is the ambition of the Europe 2020 programme, it is important to understand that this direction will be determined by policy, not by markets (or at least not with the required intensity).[11] This is not about picking specific technologies or firms, but considering the portfolio of different investments, direct and indirect, that must be made to actively influence production and consumption patterns in order to reverse the energy- and materials-intensive model inherited from mass-production—in other words, effecting a complete redirection of the entire economy. The mass-production revolution required such directionality, and Green Growth is an option in which ICT can be used to enable convergent innovations across the entire economy today. However, this revolution requires a firm push from both the supply and demand sides: supply in the sense of direct mission-oriented investment in research, development, and tax subsidies that make it clearly profitable to invest in renewable energies, special materials, conservation, recyclability, productivity of resources, and so on; demand in terms of fostering changes in consumer behaviour and stimulating the development of lagging countries,

[11] In 2010 the European Commission announced their growth strategy for the following decade, under the heading of 'Europe 2020'. The three key aims of this strategy are that growth should be 'smart, sustainable, and inclusive' (see European Commission, 2010).

where markets for appropriate technology for infrastructure and capital goods can become part of a positive-sum game with the advanced world. Given that advanced countries cannot compete in terms of cheap labour, low-value products, or services, it is essential to define—and even create—high-quality, high-value demand areas.

- *Finance: quality not quantity.* While it is often assumed that innovation requires finance, the nature of this financing is often overlooked. With its increasingly short-term goals, the venture capital (VC) model can play only a limited role. Growth-driving innovation requires long-term committed capital. Public development banks are financing up to 34 per cent of worldwide renewable energy projects, compared to only 19 per cent for corporate actors (Mazzucato and Penna, 2014). It is important for finance policy to consider the different forms of patient long-term committed capital that are required, and also to accept the high failure rates that this might entail. Equally, when public funds take the form of public venture capital, it should be possible to benefit from the upside, so that losses can be covered and the next round supported via a 'revolving fund'. This would also render investment more 'social', as called for by Keynes: 'Socialization of investments helps to replan the environment of our daily life... Not only shall we come to possess these excellent [technological] things, but... we can hope to keep employment good for many years to come' (Keynes, 1980 [1946], 270).
- *Definancialization.* Achieving this goal requires a dual approach. The total value added of the non-financial sector must be increased by inducing finance to lend not to itself but to value-creating projects in the real economy. At the same time, we need to incentivize firms to reinvest profits in areas like human capital, equipment, software, and R&D, rather than speculative areas that only aim to boost share prices (such as share buybacks), or at least find a way to limit those speculative practices.
- *Regulation for shifting relative profitability.* While it is often argued that EU businesses are impeded by red tape, such claims miss the key point, emphasized by Polanyi (1944), that markets are always regulated, being deeply embedded in political processes. The question is how they are regulated and for what purpose. Intelligent regulation (including taxation) can shift profitability towards innovation-led growth that serves the public good, which in turn can stimulate further innovation. For example, a rule that requires returning a product on disposal (as the EU directive on electrical and electronic goods does—see European Union, 2012)—stimulates redesign for ease of disassembly and recycling, rewarding the most successful designs and reducing the cost to the community. Planning rules in cities and environmental regulation of the construction industry to

minimize energy consumption and maintenance will enhance the value of properties and reduce running costs. Restricting industrial effluents can encourage 'industrial symbiosis' (Chertow, 2000), where one industry profits from using the by-products of another. Once there is a stable and consistent direction—as Green Growth can provide—regulation and innovation will tend to converge along a clear trajectory and the policy-maker's criteria for regulating will tend to coincide with those of the business strategist for innovating.

- *Taxation.* Crucial in the design of an innovation and growth strategy will be a radical and bold restructuring of the tax system. Tax structures must reward long-run investments—in labour and areas like R&D—rather than quick trades that are geared at value extraction and which lead to asset inflation.[12] If well designed, a modernized tax system would discourage casino-type financial activities, radically decrease global tax avoidance, penalize resource use while reducing the tax on labour, and favour patient capital as well as investment in the real economy. Such an innovation-oriented tax system is likely to include a 'financial transactions' tax, an increased capital gains tax (decreasing the longer the asset is held), and a shift from taxing salaries to taxing energy and materials use, as well as policies that focus on stimulating R&D and human capital formation.

The ideas behind the policies formulated since the financial collapse of 2007–8 have not contributed greatly to reviving the European economies, restoring well-paid and sufficient employment levels, or reversing income polarization. The time has come for a serious rethinking of both the policies and the premises behind them. Understanding innovation as the driver of growth, employment, and well-being, and recognising the essential role of the State in the innovation process, in creating new markets, in stimulating and giving direction to innovation, are essential foundations for such a strategic redesign. In order for there to be a real chance of success, growth policy *must become* innovation policy.

Acknowledgements

The authors wish to thank Tamsin Murray-Leach for editorial assistance on this chapter and Caetano Penna and the book editors for their useful comments. The usual disclaimers apply. Mazzucato acknowledges support from INET grant 1200037.

[12] Interestingly, the Netherlands has recently introduced an innovative R&D tax credit that, rather than targeting research income (which is easy to fudge), targets research labour (Lockshin and Mohnen, 2012)—going precisely in the direction advocated here and recognizing human capital as the sort of investment that should be encouraged.

References

Acemoglu, D. (2002). 'Technical Change, Inequality and the Labour Market', *Journal of Economic Literature*, 40(1): 7–72.

Alessandri, P., and Haldane, A. (2009). *Banking on the State*, London: Bank of England.

Amsden, A. H. (1989). *Asia's Next Giant: South Korea and Late Industrialization*, New York: Oxford University Press.

Berggren, C., and Laestadius, S. (2003). 'Co-development and Composite Clusters: The Secular Strength of Nordic Telecommunications', *Industrial and Corporate Change*, 12 (1): 91–114.

Block, F. L., and Keller, M. R. (eds) (2011). *State of Innovation: The U.S. Government's Role in Technology Development*, Boulder, Colo.: Paradigm Publishers.

Block, F. L., and Keller, M. R. (2012). 'Explaining the Transformation in the US Innovation System: The Impact of a Small Government Program', *Socio-Economic Review* (30 Sept.): 1–28 (doi:10.1093/ser/mws021), <http://ser.oxfordjournals.org/content/early/2012/09/30/ser.mws021.full.pdf+html>, accessed Feb 2013.

Boyer, R. (2000). 'Is a Finance-Led Growth Regime a Viable Alternative to Fordism? A Preliminary Analysis', *Economy and Society*, 29(1): 111–45.

Breakthrough Institute (2010). *Where Good Technologies Come From: Case Studies in American Innovation*, Oakland, Calif.: Breakthrough Institute, Dec., <http://thebreakthrough.org/blog/Case%20Studies%20in%20American%20Innovation%20report.pdf>, accessed Jan. 2013.

Broeze, F. J. A. (1982). 'The International Diffusion of Ocean Steam Navigation: The Myth of the Retardation of Netherlands Steam Navigation to the East Indies', *Economisch-en Sociaal-Historisch Jaarbook*, 45: 77–95.

Brynjolfsson, E., and McAfee, A. (2011). *Race Against the Machine: How the Digital Revolution is Accelerating Innovation, Driving Productivity, and Irreversibly Transforming Employment and the Economy*, Digital Frontier Press.

Chertow, M. R. (2000). 'Industrial Symbiosis: Literature and Taxonomy', *Annual Review of Energy and the Environment*, 25(1): 313–37.

Cohen, E. W. (1941). *The Growth of the British Civil Service, 1780–1939*, London: Allen & Unwin.

Christensen C. M., and van Bever, D. (2014). 'The Capitalist's Dilemma', *Harvard Business Review*, 92(6), 60–8.

David, P. A. (2004). 'Understanding the Emergence of Open Science Institutions: Functionalist Economics in Historical Context', *Industrial and Corporate Change*, 13 (4): 571–89.

Derthick, M., and Quirk, P. J. (1985). *The Politics of Deregulation*, Washington, DC: Brookings Institution Press.

Dore, R. (2008). 'Financialization of the Global Economy', *Industrial and Corporate Change*, 17(6): 1097–1112.

Dosi, G. (2005). *Statistical Regularities in the Evolution of Industries: A Guide through Some Evidence and Challenges for the Theory*, LEM Papers Series 2005/17, Pisa: Laboratory of Economics and Management (LEM), Sant'Anna School of Advanced Studies.

Dosi, G., and Lovallo, D. (2007). 'Rational Entrepreneurs or Optimistic Martyrs? Some Considerations on Technological Regimes, Corporate Entries, and the Evolutionary Role of Decision Biases', in Raghu Garud, Praveen Rattan Nayyar, and Zur Baruch Shapira (eds), *Technological Innovation: Oversights and Foresights*, Cambridge: Cambridge University Press, 41–68.

Dosi, G., and Nelson, R. (1994). 'An Introduction to Evolutionary Theories in Economics', *Journal of Evolutionary Economics*, 4(3): 153–72.

Dosi, G., Freeman, C., Nelson R. R., Silverberg, G., and Soete, L. (eds) (1988). *Technical Change and Economic Theory*, London: Pinter.

Dosi, G., Llerena, P., and Labini, M. S. (2006). 'The Relationships between Science, Technologies and their Industrial Exploitation: An Illustration through the Myths and Realities of the So-Called "European Paradox"', *Research Policy*, 35(10): 1450–64.

Doukas, J. (2013). 'Quantitative Easing Does Not Address the Fundamental Problems Underpinning Struggling Western Economies', blog post, LSE EUROPP (European Politics and Policy), <http://blogs.lse.ac.uk/europpblog/2013/03/04/quantitative-easing-problems>, accessed June 2013.

Ellen MacArthur Foundation (2012/2013). *Towards the Circular Economy: Economic and Business Rationale for an Accelerated Transition*, Circular Economy Report, vols 1–2, London: Ellen MacArthur Foundation.

European Commission (2010). *Communication from the Commission: Europe 2020; A European Strategy for Smart, Sustainable and Inclusive Growth*, 3 Mar., COM(2010) 2002 final, http://eur-lex.europa.eu/LexUriServ/LexUriServ.do?uri=COM:2010:2020:FIN:EN:PDF, accessed June 2013.

European Commission (2011). *Social Situation Monitor*, EU-SILC 2005–2011, <http://ec.europa.eu/social/main.jsp?catId=1050&intPageId=1871&langId=en>, accessed Apr. 2014.

European Union (2012). 'Directive 2012/19/EU of the European Parliament and of the Council of 4 July 2012 on Waste Electrical and Electronic Equipment (WEEE)', *Official Journal of the European Union*, L197 (24 July), <http://eur-lex.europa.eu/legal-content/EN/TXT/?uri=OJ:L:2012:197:TOC>, accessed Apr. 2014.

Foray, D., Mowery, D. C., and Nelson, R. R. (2012). 'Public R&D and Social Challenges: What Lessons from Mission R&D Programs?', *Research Policy*, 41(10) (Dec.): 1697–1702.

FRED (Federal Reserve Economic Data, Federal Reserve Bank of St Louis) (2013). *All Employment: Manufacturing, 1950–2013* Washington, DC: US Department of Labor: Bureau of Labor Statistics, <http://research.stlouisfed.org/fred2/graph/?s%5B1%5D%5Bid%5D=MANEMP>, accessed Nov. 2014.

Fredriksen, K. (2012). *Income Inequality in the European Union*, OECD Economics Department Working Papers, 952, Paris: OECD Publishing.

Freeman C. (1994). 'Innovation and Growth', in M. Dodgson and R. Rothwell (eds), *Handbook of Industrial Innovation*, Aldershot: Elgar, 78–93.

Freeman, C. (2008). *Systems of Innovation: Selected Essays in Evolutionary Economics*, Cheltenham: Elgar.

Freeman, C., and Perez, C. (1988). 'Structural Crises of Adjustment: Business Cycles and Investment Behaviour', in G. Dosi et al. (eds), *Technical Change and Economic Theory*, London: Pinter, 38–66.

Freeman, C., and Soete, L. (1994). *Work for All or Mass Unemployment: Computerised Technical Change into the 21st Century*, London: Pinter.

Geels, F. W. (2002). 'Technological Transitions as Evolutionary Reconfiguration Processes: A Multi-Level Perspective and Case Study', *Research Policy*, 31: 1257–74.

Gordon, R. (2012). *Is U.S Economic Growth Over? Faltering Innovation Confronts the Six Headwinds*, Working Paper, 18315, National Bureau of Economic Research, Aug., <http://www.nber.org/papers/wl8315>, accessed June 2013.

Grunwald, M. (2012). *The New New Deal: The Hidden Story of Change in the Obama Era*, New York: Simon & Schuster.

Haldane, A. G. (2011). 'The Short Long', speech at the 29th Société Universitaire Européene de Recherches Financières Colloquium: New Paradigms in Money and Finance? Brussels.

Hobsbawm, E. (1994). *The Age of Extremes*, London: Michael Joseph.

Juncker, J.-C., Tusk, D., Dijsselbloem, J., Draghi, M. and Schulz, M. (2015). Completing Europe's Economic and Monetary Union. Brussels: 22.10.2015, HYPERLINK <http://ec.europa.eu/priorities/economic-monetary-union/docs/5-presidents-report_en.pdf%20accessed%2020%20July%202015> <http://ec.europa.eu/priorities/economic-monetary-union/docs/5-presidents-report_en.pdf>, accessed 20 July 2015.

Kaldor, N. (1966). *Causes of the Slow Growth in the United Kingdom*, Cambridge: Cambridge University Press.

Keynes, J. M. (1926). *The End of Laissez-Faire*, London: L & V Woolf.

Keynes, J. M. (1931). 'The Grand Slump of 1930', in *Essays in Persuasion: The Collected Writings of John Maynard Keynes*, London: Macmillan for the Royal Economic Society, 1972, xi. 126–49.

Keynes, J. M. (1936). *The General Theory of Employment, Interest and Money*, New York: Harcourt, Brace & Co.

Keynes, J. M. (1938). 'Private letter to Franklin Delano Roosevelt', 1 Feb., in D. E. Moggridge, *Maynard Keynes: An Economist's Biography*, London: Routledge, 1992.

Keynes, J. M. (1980). 'Activities 1940–46: Shaping the Post-War World: Employment and Commodities', in *Collected Works*, xxvii, ed. D. Moggridge, London: Macmillan.

Knight, F. (1921). *Risk, Uncertainty and Profit*, New York: Augustus M. Kelley.

Krippner, G. R. (2005). 'The Financialization of the American Economy', *Socio-Economic Review*, 3(2): 173–208.

Lazonick, W. (2009). *Sustainable Prosperity in the New Economy? Business Organization and High-Tech Employment in the United States*, Kalamazoo, MI: WE Upjohn Institute.

Lazonick, W. (2014). 'Profits without Prosperity', *Harvard Business Review*, 92(9): 46–55.

Lazonick, W., and Mazzucato, M. (2013). 'Risks and Rewards in the Innovation–Inequality Relationship', *Industrial and Corporate Change*, ed. M. Mazzucato, special issue on 'Finance, Innovation, and Growth', 22(4): 1093–1128.

Lazonick, W., and O'Sullivan, M. (2000). 'Maximizing Shareholder Value: A New Ideology for Corporate Governance', *Economy and Society*, 29(1): 13–35.

Lazonick, W., and Tulum, O. (2011). 'US Biopharmaceutical Finance and the Sustainability of the Biotech Business Model', *Research Policy*, 40(9): 1170–87.

Lipsey, R., Carlaw, K. I., and Bekhar, C. T. (2005). *Economic Transformations: General Purpose Technologies and Long-Term Economic Growth*, Oxford: Oxford University Press.

Locke, R. M., and Wellhausen, R. (eds) (2013). 'A preview of the MIT Production in the Innovation Economy Report', 22 Feb., <http://web.mit.edu/press/images/documents/pie-report.pdf>, accessed Feb. 2013.

Lockshin, B., and Mohnen, P. (2012). *Do R&D Tax Incentives Lead to Higher Wages for R&D Workers? Evidence from the Netherlands*, Maastricht Economic and Social Research Institute on Innovation and Technology (UNU-Merit), Working Paper, 2012-058, 18 July; revised version of Working Paper, 2008-034.

Lombardi, D. R., and Laybourn, P. (2012). 'Redefining Industrial Symbiosis', *Journal of Industrial Ecology*, 16(1): 28–37.

Lundvall, B.-Å. (ed.) (1992). *National Innovation Systems: Towards a Theory of Innovation and Interactive Learning*, London: Pinter Publishers.

Lundvall, B.-Å., Johnson, B., Andersen, E. S., and Dalum, B. (2002). 'National Systems of Production, Innovation and Competence Building', *Research Policy*, 31: 213–31.

McDonough, W., and Braungart, M. (2002). *Cradle to Cradle: Remaking the Way we Make Things*, New York: North Point Press.

Mazzucato, M. (2012). 'Finance and Innovation: Reforming a Dysfunctional System', FINNOV Policy Brief 3, <www.finnov-fp7.eu/sites/default/files/FINNOV%20Policy%20Brief%20Feb%201%202012.pdf>, accessed Apr. 2014.

Mazzucato, M. (2013a). *The Entrepreneurial State: Debunking Private vs. Public Sector Myths*, London: Anthem Press.

Mazzucato, M. (2013b). 'Finance, Innovation and Growth: Finance for Creative Destruction vs. Destructive Creation', special issue of *Industrial and Corporate Change*, ed. M. Mazzucato, 22(4): 869–901.

Mazzucato, M. (2015). 'Building the Entrepreneurial State', Levy Institute Working Paper No. 824, January.

Mazzucato, M., and Penna, C. (2014). 'Beyond Market Failures: The Role of State Investment Banks in the Economy', unpublished working paper.

Mazzucato, M., and Shipman, A. (2014). 'Accounting for Productive Investment and Value Creation', *Industrial and Corporate Change* (14 Jan.), <http://icc.oxfordjournals.org/content/early/2014/01/13/icc.dtt037.short?rss=1>, accessed May 2014.

Mowery, D. C. (2010). 'Military R&D and Innovation', in B. H. Hall and N. Rosenberg (eds), *Handbook of the Economics of Innovation*, Amsterdam: North-Holland, ch. 29.

National Endowment for Science Technology and the Arts (2009). *The Vital 6 Per Cent: How High-Growth Innovative Businesses Generate Prosperity and Jobs*, Research Summary, London: NESTA.

ONS (2013). 'Labour Market Statistics 1978–2013', <www.ons.gov.uk>.

O'Sullivan, M. (2005). 'Finance and Innovation', in Jan Fagerberg, D. Mowery, and R. R. Nelson (eds), *The Oxford Handbook of Innovation*, Oxford: Oxford University Press, ch. 9.

Pavitt, K. (1984). 'Sectoral Patterns of Technical Change: Towards a Taxonomy and a Theory', *Research Policy*, 13(6): 343–73.

Perez, C. (2002). *Technological Revolutions and Financial Capital: The Dynamics of Bubbles and Golden Ages,* Cheltenham: Edward Elgar.

Perez, C. (2009). 'The Double Bubble at the Turn of the Century: Technological Roots and Structural Implications', *Cambridge Journal of Economics*, 33(4): 779–805.

Perez, C. (2013a). 'Financial Bubbles, Crises and the Role of Government in Unleashing Golden Ages', in A. Pyka and H. P. Burghof (eds), *Innovation and Finance,* London: Routledge, 11–25.

Perez, C. (2013b). 'Unleashing a Golden Age After the Financial Collapse: Drawing Lessons from History', *Environmental Innovation and Societal Transitions,* 6: 9–23.

Piketty, T. (2014). *Capital in the 21st Century,* Cambridge, Mass.: Harvard University Press.

Piketty, T., and Saez, E. (2010). 'Income Inequality in the United States, 1913–1998', *Quarterly Journal of Economics,* 115(1): 1–39 (update of 2003), <http://elsa.berkeley.edu/~saez/#income>, accessed Mar. 2013.

Polanyi, K. (2001). *The Great Transformation: The Political and Economic Origins of our Time,* Boston: Beacon (orig. 1944).

Rajan, R. (2014). 'World Leaders Must Re-Examine Global Rules of Monetary Game', *Gulf Times,* 29 Apr. (based on a speech given at the Brookings Institution, 12 Apr.), <www.gulf-times.com/opinion/189/details/390162/world-leaders-must-re-examine-global-rules-of-monetary-game>, accessed 2014.

Ricardo, D. (1817). *On the Principles of Political Economy and Taxation,* London: John Murray.

Rockström, J., et al. (2009). 'Planetary Boundaries: Exploring the Safe Operating Space for Humanity', *Ecology and Society,* 14(2): 32, <www.ecologyandsociety.org/vol14/iss2/art32>.

Romer, P. M. (1994). 'The Origins of Endogenous Growth', *Journal of Economic Perspectives,* 8(1): 3–22.

Rosenberg, N. (1983). *Inside the Black Box: Technology and Economics,* Cambridge: Cambridge University Press.

Sachs, J. (2014). 'We Need a Sixth Wave of Sustainable and Green Technologies', Green Week Conference 2014, European Commission, Brussels, 3 June, <http://cordis.europa.eu/news/rcn/36601_en.html>, accessed Apr. 2014.

Schumpeter, J. (1934). *The Theory of Economic Development,* Harvard Economic Studies, Cambridge, Mass.: Harvard College (orig. 1912).

Schumpeter, J. A. (1982). *Business Cycles,* Philadelphia: Porcupine Press (orig. 1939).

Stirling, A. (2009). *Direction, Distribution and Diversity: Pluralising Progress in Innovation, Sustainability and Development,* STEPS Working Paper, 32, Brighton, STEPS Centre.

Storey, D. J. (2006). 'Evaluating SME Policies and Programmes: Technical and Political Dimensions', in M. Casson (ed.), *The Oxford Handbook of Entrepreneurship,* Oxford: Oxford University Press, 248–78.

Summers, L. (2013). 'Why Stagnation Might Prove to Be the New Normal', *Financial Times,* 15 Dec., <www.ft.com/cms/s/2/87cb15ea-5d1a-11e3a55800144feabdc0.html#axzz35Bf2xOBb>, accessed Apr. 2014.

Van Ark, B. (2001). *The Renewal of the Old Economy: An International Comparative Perspective,* Paris: OECD.

Varoufakis, Y., Holland, S., and Galbraith, J. K. (2013). 'A Modest Proposal for Resolving the Euro Crisis, Version 4.0', <http://varoufakis.files.wordpress.com/2013/07/a-modest-proposal-for-resolving-the-eurozone-crisis-version-4-0-final1.pdf>, accessed May 2014.

Verdoorn, J. P. (1949). 'On the Factors Determining the Growth of Labor Productivity', in L. Pasinetti (ed.), *Italian Economic Papers*, Oxford: Oxford University Press, 1993), ii. 59–68.

Wade, R. (1990). *Governing the Market: Economic Theory and the Role of Government in East Asian Industrialization*, Princeton: Princeton University Press.

Womack, J. P., Jones, D. T., and Roos, D. (1990). *The Machine that Changed the World*, New York: Rawson Associates.

Zenghelis, D. (2011). *A Macroeconomic Plan for a Green Recovery*, Centre for Climate Change Economics and Policy, Grantham Research Institute on Climate Change and the Environment, Policy Paper, Jan, <http://www2.lse.ac.uk/GranthamInstitute/publicatio<ns/Policy/docs/PP_macroeconomic-green-recovery_Jan11.pdf>, accessed May 2014.

Index

Printed and bound by CPI Group (UK) Ltd, Croydon, CR0 4YY